Gordon McComb's Tips & Techniques for the Electronics Hobbyist

Gordon McComb

TAB TAB BOOKS
Blue Ridge Summit, PA

FIRST EDITION
FIRST PRINTING

Published by TAB Books.
TAB Books is a division of McGraw-Hill, Inc.

Library of Congress Cataloging-in-Publication Data

McComb, Gordon.
[Tips & techniques for the electronics hobbyist]
Gordon McComb's tips & techniques for the electronics hobbyist / by Gordon McComb.
p. cm.
Includes bibliographical references and index.
ISBN 0-8306-7485-3 (h) ISBN 0-8306-3485-1 (p)
1. Electronics—Amateurs' manuals. I. Title. II. Title: Gordon McComb's tips and techniques for the electronics hobbyist
TK9965.M36 1991
621.381—dc20 90-28653
CIP

TAB Books offers software for sale. For information and a catalog, please contact TAB Software Department, Blue Ridge Summit, PA 17294-0850.

Questions regarding the content of this book should be addressed to:

Reader Inquiry Branch
TAB Books
Blue Ridge Summit, PA 17294-0850

Acquisitions Editor: Roland S. Phelps
Technical Editor: Lisa A. Doyle
Production: Katherine G. Brown
Book Design Jaclyn J. Boone

Contents

Introduction

SAVE TIME, MONEY, AND FRUSTRATION. THAT'S THE THEME OF THIS BOOK, which offers you, in one handy volume, all the basic information for experimenting with hobby electronics. *Gordon McComb's Tips and Techniques for the Electronics Hobbyist* is a compendium of useful information for the general electronics enthusiast. In many ways, it is a "best of" book, detailing the most noteworthy topics of interest to the electronics hobbyist. It does not dwell heavily in any one area of the electronics field but provides lucid, to-the-point facts—answers to questions most often asked.

The topics addressed in this book include:

- Setting up shop: the best place for electronics work, including lighting, tools, and toolboxes, proper environment and atmosphere.
- The basics of test equipment: all about test equipment used by electronics enthusiasts including multimeters, scopes, logic probes, AF signal injectors, signal generators, and counters.
- How to get the most from your test equipment.
- Where to buy components and where to find the best deals. Plus, standard electronic components you can't live without and need for your basic stock (this section saves you money by helping you to stock up only on what you need the most).
- Basic schematic symbols and their meanings, including how to read a schematic.

- Proper identification of components (ICs, transistors, resistors, capacitors, inductors, transformers, and diodes) is a must for even basic project building or repair.
- How to use direct etch, photo transfer, and dry transfer techniques to build your own printed circuit boards. Also included are etching secrets of the pros.
- How to repair damaged printed circuit boards.
- Using and understanding electronic formulas including those for Ohm's law, power, series and parallel resistor networks, series and parallel capacitor networks, and the 555 timer IC.
- Benchtop projects: money-saving test equipment you can build yourself for your own benchtop, including a frequency counter, power supplies, a logic probe, a logic pulser, and many more.

Who This Book is For

This book was written for a wide variety of readers. It contains general information on electronics practice, important formulas, tips on how to identify components, and more. You can use it as a sourcebook of ideas, as a textbook on electronics techniques and procedures, or as a databook on electronics formulas, functions, and components.

This book requires a minimal electronics background. Projects appearing in chapter 20 are built using standard circuit-making techniques, all of which are detailed in previous chapters. You need to know how to read a schematic diagram only if you care to build one of the projects in chapter 20. For your convenience, you'll find a chapter on schematic symbols and fundamentals of schematics reading (chapter 7).

If you've been looking for a single, concise handbook of general electronics information, this book will serve to answer all of your questions. The 20 chapters in *Gordon McComb's Tips and Techniques for the Electronics Hobbyist* are devoted to everything electronic—from how to set up a working electronics bench, to how to build and use accurate test equipment, to how to manufacture your own printed circuit boards, how to properly use a wire-wrap tool, and much more.

If you have some experience in electronics, you can skip around and read only those chapters that provide the information you're looking for. The chapters are very much stand-alone modules. This allows you to pick and choose, using your time to its best advantage. However, if you're new to electronics, you should take a more pedestrian approach and read as much of this book as possible. In this way, you'll get a thorough understanding of the art and science of electronics, allowing you to get a lot more out of this rewarding hobby.

How to Use This Book

The Electronics Hobbyist's Benchtop Companion is divided into 20 chapters. Here's a quick rundown of each chapter.

Chapter 1, "Setting Up Shop," explains a place to work, good lighting, tools and toolboxes, proper environment and atmosphere, and fulfilling basic needs.

Chapter 2, "Filling the Toolbox," covers what you need to complete the benchtop, such as tools and supplies.

Chapter 3, "Buying and Using Test Equipment," is all about meters (digital and analog), oscilloscopes, logic probes, AF signal injectors, signal generators, and frequency counters.

Chapter 4, "Buying Components," gives guidance on where to buy (the advantages and disadvantages of local electronic stores, store chains (like Radio Shack), mail-order companies, and swap meets, saving money and stocking up, and parts and components you can't do without.

Chapter 5, "Electronics and Safety," emphasizes that safety comes first, including general safety requirements, working with ac, working with high voltage, physical safety (cuts, sharp edges, etching fluid, etc.), and soldering safety.

Chapter 6, "Eliminating Static Electricity," discusses what static electricity is and why it's a problem, safeguards you can readily take, using an antistatic wriststrap, using antistatic pads and shields, proper clothing, and grounding tools.

Chapter 7, "Schematics and Electronic Symbols," lists basic schematic symbols and their meanings and discusses reading a schematic and identifying circuit input, output, and voltage sources.

Chapter 8, "All About Batteries," shows how to effectively choose and use batteries including alkaline, heavy-duty flashlight, nickel-cadmium, lead-acid, and "gel-cell."

Chapter 9, "Component Identification," explains proper identification of components as a must for even basic project building or repair: identifying resistors, capacitors, integrated circuits, transistors, diodes, inductors, and transformers.

Chapter 10, "Common Component Pinout Diagrams," covers pinout diagram and lead reference for many TTL, CMOS, and linear ICs.

Chapter 11, "The Science of Soldering," explains the how-to's of soldering, desoldering, inspection, soldering ICs and other components, and tips from the pros.

Chapter 12, "Using Experimenter's Breadboards and Solderboards," discusses what an experimenter's breadboard is and how to use one.

Chapter 13, "Circuit Board Layout Techniques," covers how to use direct etch, photo transfer, and dry transfer PCB-making techniques and the pros and cons of each.

Chapter 14, "Techniques in Printed-Circuit-Board Etching," describes etching the board after laying out the master and obtaining optimum results with the least amount of time and money.

Chapter 15, "Wire wrapping," covers techniques of wire wrapping, with an emphasis on proper tools and planning.

Chapter 16, "Repairing Electronic Circuit Boards," details how to repair broken PCBs by resoldering, replacing, or repairing broken traces, repairing broken board substrate, adding new traces, and replacing components.

Chapter 17, "Using Electronic Formulas," contains the basic formulas for Ohm's law, power, series and parallel resistor networks, series and parallel capacitor networks, and miscellaneous electronic formulas.

Chapter 18, "Circuit Grab Bag," contains many of the most frequently used circuits and circuit fragments (with design equations for tailoring circuit for specific application): 555 monostable, 555 astable, op amp (such as an amp, signal generator, or filter), crystal timebase, LED/lamp/relay driver, switch debounce circuits (TTL and CMOS), and tone decoder/PLL.

Chapter 19, "Interfacing Logic Families and ICs," discusses TTL-to-CMOS level translators, CMOS-to-TTL level translators, interfacing from op amp to CMOS and TTL, and opto-isolator circuits for buffering and level shifting.

Chapter 20, "Benchtop Projects," includes a variety of useful testing equipment you can make for your benchtop. Many of these projects are simple, containing only one to three IC circuits.

In addition, the appendices provide sources for electronics parts and supplies and a list of magazines and books for further reading. All projects used in this book have been thoroughly tested in prototype form.

Projects

1

Setting up shop

IN JUST ABOUT EVERY 1950s SCIENCE FICTION THRILLER, THERE'S A SCIENTIST working in an elaborate laboratory. This lab was built in the outskirts of town, tucked away in the hollow of a canyon or perched high atop a mountain, overlooking the populace at the nearby city.

This laboratory would be equipped with all the latest electronic gear: interocitors, flux capacitors, infinitely indexed memory banks—you name it, they had it. Surrounding this impressive collection of science and technology was an electrified fence and, for good measure, an armed guard to ensure that all remained well (and that poor guard was always the first one to get knocked off by the monster).

For hobby electronics work, you don't need such lofty quarters. A corner of a bedroom, a small slice of the family room, or a section of the garage should be perfect for your needs. The prime ingredients for the electronics laboratory are:

- A comfortable place to work, with table and chair.
- Good lighting.
- Ample electrical outlets, with at least 15-amp service.
- Tools and toolboxes, nearby and readily accessible.
- Peace and quiet.

Let's quickly review what you need to set up shop for electronics tinkering. I also discuss proven workshop layouts you readily duplicate in your own home and detail a handy electronics bench you can build for your workshop.

A Place to Work

Before setting up shop, consider the best place in your house for building electronics projects. The garage is an ideal setting because it affords you the freedom to solder, hammer, and etch without worrying about soiling the new carpet. You don't need much space—about 3 by 4 feet ought to do it—so you can often build an electronics work area and still park the car and have room for household storage. Obviously, the larger the garage, the better. A two- or three-car garage affords greater room to stretch out.

A room in the house is a good second choice, but only if it conforms to some basic requirements. I've found that when working in a carpeted room, it's best to spread another carpet or some protective cover over the floor, for example, an antistatic mat (covered in chapter 6, "Eliminating Static Electricity"). When the throw rug or cover fills with solder bits and little pieces of wire and component leads, take it outside, beat it with a broom handle, and it's as good as new.

A bedroom, den, or family room is an acceptable location for your electronics lab, but clear away a corner or section of the room for the dedicated use of electronics hacking. Odds are that you'll need to leave projects overnight—or even longer periods of time—and it's important that your work be undisturbed.

If your work area is in a main area of the house, you might want (or need) to tuck it away when you're not experimenting. Working with electronics can be messy, especially if you're in the middle of troubleshooting your latest project. A folding panel works wonders to hide your work area, especially if the desk is situated in the corner of the room, as shown in Fig. 1-1.

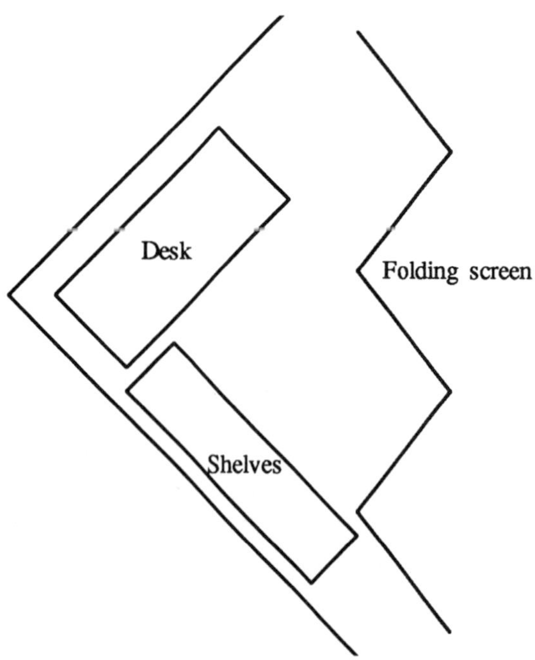

Fig. 1-1. Use a folding panel to hide your electronics bench when not in use.

If you're working in a bedroom or den, you might consider placing the electronics bench in the closet. Close the closet doors and no one would know you're building a laser-beam voice communication system, or whatever else you're currently working on.

No matter where you set up shop, consider heat and humidity. If your work area is at all uncomfortable to you, it is not suitable for electronics work. Extremes in heat or cold can have a profound effect on your electronics circuits.

If you're working in a garage, attic, or basement, consider adding insulation if the area doesn't already have it. Rolls of fiberglass insulation are relatively inexpensive, and installation requires little more than a staple gun. At most, it's an afternoon job. Be sure to follow the installation directions carefully, and wear gloves, eye protection, and a respirator.

Basements and garages pose problems of excessive moisture. If your basement is at or below the water table level, water could accumulate on the floor. For safety reasons, *never* work in an area where the floor is wet or even slightly damp. This is especially true of electronics work, where you might need to handle potentially lethal voltages of electricity.

When working in the garage, keep your electronics bench away from doors and other openings. This prevents outside moisture from ruining your projects. It also helps keep grass, bugs, and dust out of your circuit boards.

For best results, your workspace should be an area where your project-in-progress won't be disturbed if you have to leave it for several hours or several days, as will often be the case. The worktable should also be one that is off limits or inaccessible to young children, or at least an area that can be easily supervised. This is especially true for electronics projects that work with high voltages such as laser systems, plasma globes, or power supplies.

Your Workbench

Your workbench doesn't have to be large or elaborate. The size of the workbench is mainly determined by the type of projects you'll be doing, but for most applications, a table measuring about 2 by 3 feet is more than enough. You probably already have a small desk, table, or drafting table you can use for your electronics bench.

You can also use a door as a table surface. Build legs using 30-inch lengths of 2-by-4 lumber, as shown in Fig. 1-2. Hollow-core doors are cheaper, but the solid-core doors last longer and don't bow with age and weight. As an alternative, you can build your work surface using 3/4-inch plywood or particle board.

Many electronics technicians prefer to cover the work surface with a layer of carpeting. The carpeting acts as a cushion to protect circuit boards, cabinets, and other components. If you use a piece of carpet, get a new, clean remnant and cut it to size. The shorter the nap, the better. Closed-loop naps are better

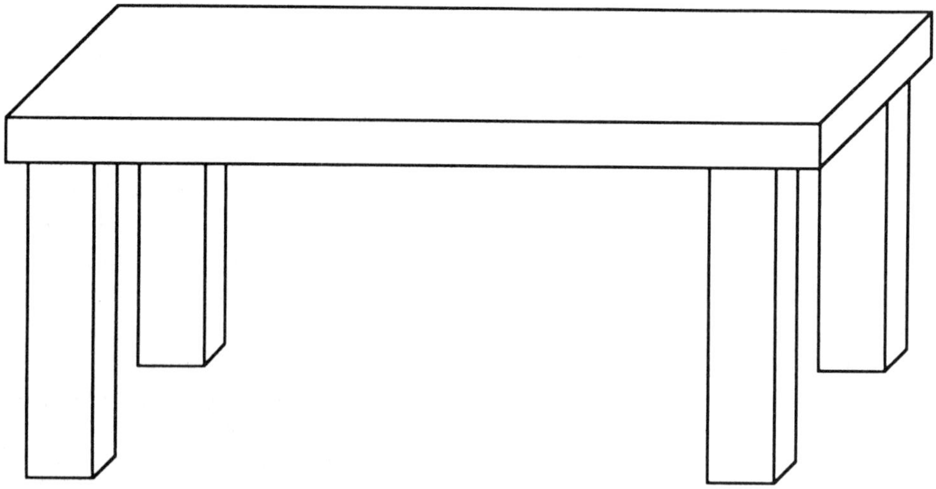

Fig. 1-2. A hollow- or solid-core door makes an excellent workbench.

than cut naps (like plush and saxony carpeting). Avoid shags at all costs. Static build-up can be a problem with some carpets. If you can, get a carpet that has been treated with an antistatic spray, or better yet, uses antistatic (metallic) threads.

You'll be crouched over the worktable for hours at a time, so a comfortable chair or stool is a must. Be sure the seat is adjusted for the height of the worktable. A poor-fitting chair causes backaches and fatigue. You can skimp and buy or build an inexpensive worktable, but if you don't already own a good chair, it should be on the top of your shopping list.

Lighting

Good lighting is a must. Both mechanical and electronic assembly require detailed work, and you need good lighting to see everything properly. Fluorescent overhead lights provide good overall illumination at low cost. Supplement overhead lights with a 60-watt or higher desk lamp placed on your worktable.

Tools and Toolboxes

You'll need a screwdriver and a few other common tools to build electronics projects. Chapter 2 is devoted to the tools of the electronics trade, so this section just covers the basic requirements and where to put those tools so you have instant access to them.

At a minimum, you need a set of screwdrivers (both flathead and Phillips), a soldering iron, wire cutters, wire strippers, and small pair of needle-nosed pliers. The soldering iron should be stored in its stand; the screwdrivers and other tools should be stashed in a small toolbox, which you can keep on the workbench. You can purchase inexpensive toolboxes for under $3.

Another approach is to use a plastic fishing tackle box. The tackle box has lots of small compartments for placing screws and other parts and includes a large section for storing your basic tools.

If your screwdrivers are not already magnetized, purchase a screwdriver magnetizer at the hardware store. When magnetized, the screwdriver holds on to screws that you remove or reinstall into your electronics projects. The magnetizer lets you magnetize and demagnetize your screwdrivers and other metal tools. Or, you can magnetize your screwdrivers by scraping the blade across the face of a large speaker magnet.

A good pair of pliers is a handy tool to have around, and you might want to purchase several, each a difference size, for different jobs. Use the pliers to loosen or tighten nuts, grommets, and plastic standoffs.

Tweezers or a pair of small, long-nosed pliers help you grasp small parts—like screws that have fallen inside the deck! A pair of regular manicuring tweezers is fine, but try to get the kind with the flat, blunt end. Tweezers with a pointed end aren't as useful.

Proper Environment

You'll have more fun working on electronic projects if your work area is neat and clean. Regularly vacuum or sweep the floor. Scrap components and wire leads are often scattered about when working on a project; gather them up and dispose of them (you should sift through the waste to look for components that are still good and lengths of bare wire that you can use for circuit jumpers). Get into the habit of cleaning up after each day, or at least after each project.

Even if the projects you build operate under battery power, you need an ac outlet for your soldering iron and electronic test gear. For ready access to electricity, purchase a multiple-outlet power strip (cost is under $15) at any home-improvement or hardware store. The strip includes its own on/off switch (usually lighted) and a circuit breaker rated for 15 or 20 amps. The better power strips also come with surge protectors.

If you're working on a concrete floor, consider adding a small rug or anti-static mat. The rug or mat will keep you warmer in the winter months and reduces the chance of electrical shock due to moisture trapped in the concrete floor.

For your own health and safety, your electronics lab should be equipped with a fan or ventilator. If the area is small, you can use a 6- or 8-inch fan, positioned so that air blows across the table and out a window or door. Larger areas might require heavy-duty fans or even a dedicated wall-mounted ventilator. Be especially wary of smoke and fumes created during soldering. These fumes are mildly toxic, and in heavy or prolonged doses, they can make you sick.

Good ventilation is also important if you use electrical cleaners, solvents, and chemicals. Advanced electronics experiments using circuit-board etching processes *absolutely* require proper ventilation or serious respiratory injury could result. Refer to chapters 5 and 14 for more information on using dangerous chemicals and circuit-board etchants.

If you have $75 to spare, consider investing in an electrostatic air cleaner. These air cleaners use high voltage to filter the air and trap smoke particles and dust.

Shelves and Cabinets

The longer you experiment in electronics, the more junk you collect. This junk is stored in—what else—junk boxes. Many a project has been built from parts scavenged solely from the junk box.

Because there are so many types of electronic components, piling all of your surplus into a single box is decidedly a bad idea. By organizing your supplies in a thoughtful and consistent manner, you can more easily find a part when you need it. You can use small plastic parts cabinets (available at most home improvement and hardware stores) to store your various goodies. Identify the contents of each drawer with a label, as shown in Fig. 1-3. You can write the label on a sticker or use a Dymo labeler if you want a more professional look.

Parts cabinets with various sizes of drawers seem well suited for storing electronic components. Put the bulky items like transformers, power resistors, and electrolytic capacitors in the big drawers, and save the smaller drawers for the low-wattage resistors, signal diodes, and transistors.

Fig. 1-3. A sample parts cabinet.

Parts you don't need ready access to can be stored in quart-sized resealable plastic bags, as shown in Fig. 1-4. You don't need to label the contents of the bag because you can readily identify what's inside. Use heavy-duty bags if the components have sharp edges or leads; this prevents the parts from poking through and ripping holes in the bags.

Store your cabinets and parts bags on shelves. If you can, install wall-mounted shelves over your workbench. The shelves might consist of two or three uprights, or standards, and a pair of brackets for each shelf you use. The brackets are available in standard shelf widths, including 6, 10, 12, and 15 inches wide. The shelf standards must be securely fastened to the wall.

If you are installing the standards over wallboard, use anchor bolts of one design or another (such as "butterfly" bolts or expansion bolts). Use large wood screws when securing the standards into the 2-by-4 wall studs.

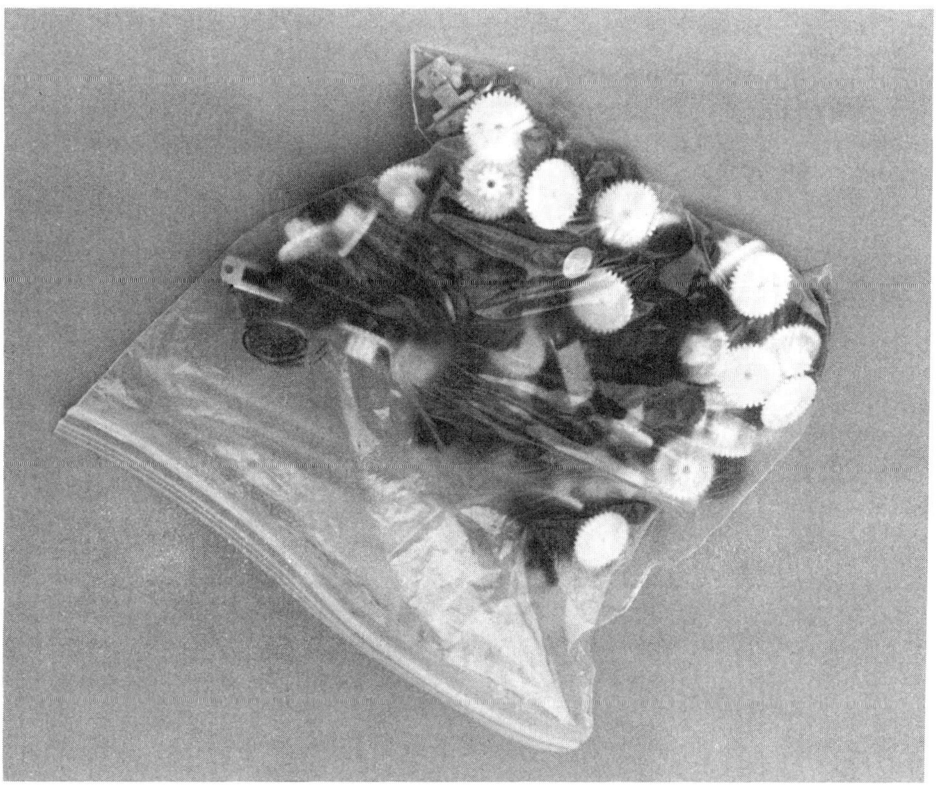

Fig. 1-4. Put assorted electronic and mechanical parts in heavy-duty resealable bags.

Sample Workshop Layouts

Your needs and the design of your room or garage will greatly determine the actual layout of your electronics work area. But here are some layout ideas you might want to consider.

Corner installation—garage

Corner installations afford the greatest amount of workspace in an area that might otherwise be wasted. If you're lucky enough to have an empty corner in your garage that's away from the furnace, woodworking areas (to avoid saw dust), and doors and openings that lead outdoors, it's a good candidate for your electronics workbench.

Figure 1-5 shows a typical corner installation in a garage, using a 3-by-4 foot drafting table. Shelves are installed on the wall over the table, and cables, test clips, and other odds and ends are attached to the corner wall using nails and wooden pegs.

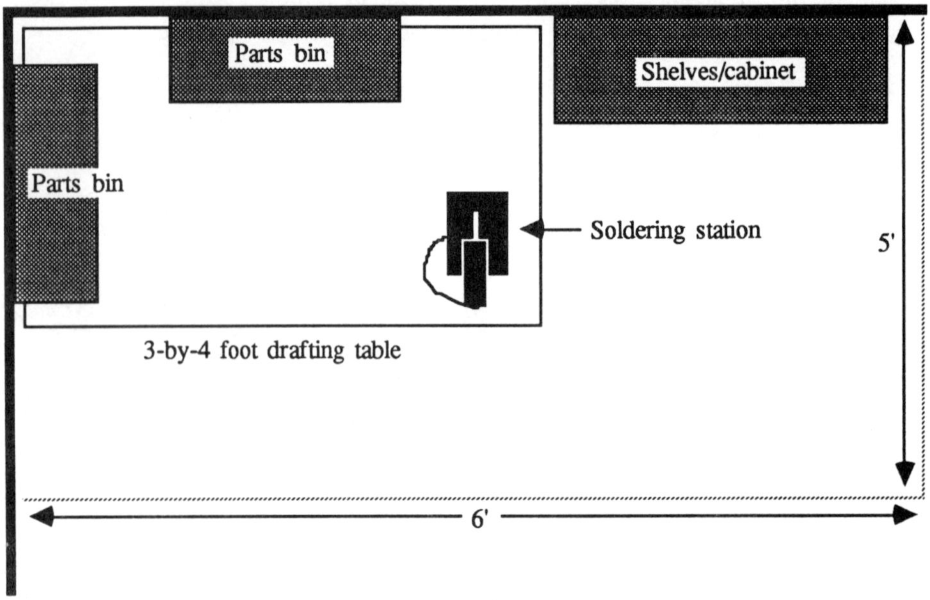

Fig. 1-5. One way to set up shop in the corner of a garage.

Wall installation—garage

Small garages might have limited space in the corners to set up an electronics workbench. If you park your car in your garage, measure the distance between the car and one wall (preferably a wall that joins to the rest of the house rather than an outside wall). If the distance is 36 inches or more, you might be able to cram a worktable against the wall.

As shown in Fig. 1-6, remember to allow enough room for the table as well as your chair and your body. If you'll be working in the garage when the car is inside, add a piece of foam or thick fabric to the back of the chair, to protect the car against scratches.

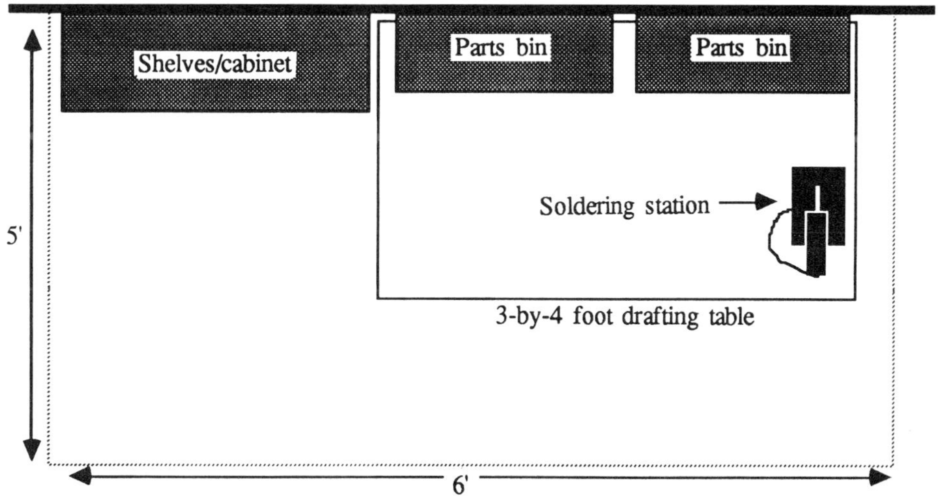

Fig. 1-6. One way to set up shop along the wall in a garage.

Corner or wall installation—bedroom or den

Use a corner of a bedroom or den as you would a corner of the garage. Place the worktable in the corner, with shelves for supplies on the wall. If you can't install shelves directly on the walls, consider a small wooden bookshelf placed to the side of the worktable.

Alternatively, you can use an empty wall in a bedroom or den for an electronics work area.

Closet installation—bedroom

Installing your electronics bench in the closet allows you to quickly close off your work in progress, which simultaneously hides the mess and protects the little ones from accidental injury. The closet should be deep enough for an ordinary study desk (about 20 inches). If you need to share the closet with clothes, be sure the clothes don't drape over the worktable, especially the soldering iron. Hot soldering irons can ignite clothes.

Build Your Own Workbench

If you don't already have a suitable desk or table for your electronics work, you can readily build one using a solid-core ash door and ordinary 2-by-4 lumber. The construction diagram for the workbench is shown in Fig. 1-7; the parts are listed in Table 1-1. The door specified is 24 inches wide, which is designed for a small doorway. The length is about seven feet, but as long as the door has a solid core, you can cut it to size if necessary.

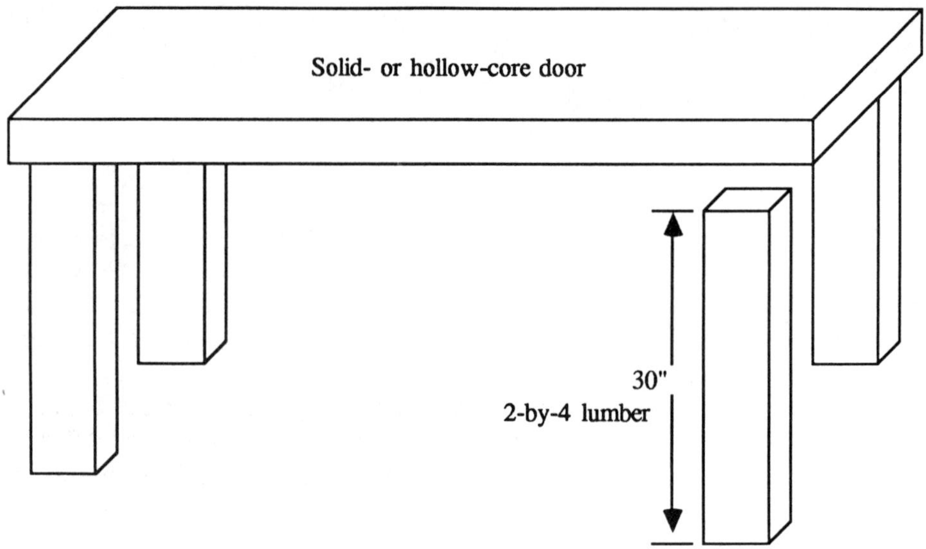

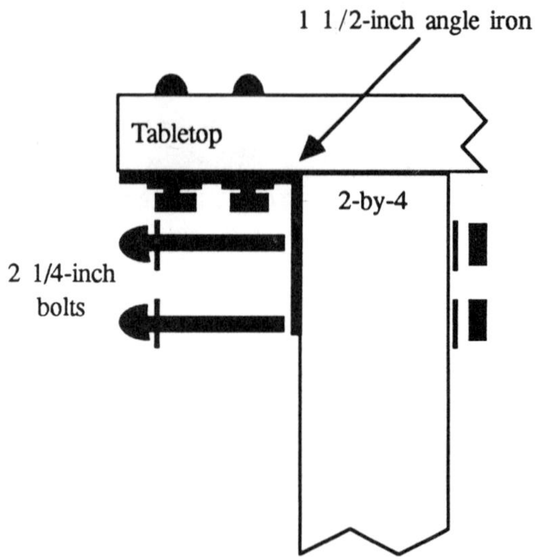

Fig. 1-7. Details on using a door and 2-by-4 lumber to construct an electronics workbench. Cost of materials (when using a hollow-core door): under $25.

Table 1-1. Parts List for Workbench

1	Solid- or hollow-core door (solid-core recommended)
4	30-inch lengths 2-by-4 lumber
4	$1\frac{1}{2}$-$\frac{1}{2}$-inch corner angle iron
16	$\frac{1}{4}$-inch 20 bolts, $2\frac{1}{4}$ inches long
16	$\frac{1}{4}$-inch 20 nuts
32	$\frac{1}{4}$-inch 20 flat washers

Start by attaching the metal 2-by-4 supports to the four corners of the door. Though the supports are designed to be attached with nails, use 1-inch by #10 wood screws for added stability. Then, cut the 2-by-4 lumber to 30 inches—longer if you want a high desk, but no shorter unless you like to sit on the floor. Be sure to cut the lumber squarely, or the legs will be wobbly and uneven. Attach the 2-by-4 legs to the metal supports, again using 1-inch by #10 wood screws.

As an alternative design, purchase or make a pair of sawhorses and secure the door to the tops of the sawhorses. You can often purchase ready-made sawhorses at home-improvement or lumber stores for $10 to $15 each. You don't need the heavy-duty kind because your electronics workbench won't be supporting any more than 40 or 50 pounds.

If you need drawers, a quick and inexpensive approach is to use prefabricated particle-board drawer units. These units are available in several different sizes and styles and come with all parts and precut wood pieces. You can assemble a typical three-drawer unit in less than an hour. The units are designed for stacking, so you can add as many drawers as you need. The cost is about $25 for an 18-inch-wide three-drawer unit. Once assembled, you can leave the particle board unfinished, or you can paint or decorate it as you like.

2

Filling the toolbox

TAKE A LONG LOOK AT THE TOOLS IN YOUR GARAGE OR WORKSHOP. YOU PROBABLY have all the implements necessary to build electronic projects. Unless your designs require a great deal of precision (and most don't), a common assortment of hand tools are all that's really required to construct all sorts of electronic goodies.

This chapter discusses the basic tools and supplies you need for experimenting with hobby electronics and how you might use them. Consider this chapter a guide only; suggestions for tools and supplies are just that—suggestions. By no means should you feel that you must own each tool mentioned in this chapter or have on hand all the parts and supplies.

Some supplies and parts might not be readily available to you, and it's up to you to consider alternatives and how to work these alternatives into the design. Ultimately, it will be your task to take a trip to the hardware store, collect various miscellaneous items, and go home to hammer out a unique creation that's all your own.

Note that this chapter details basic tools, cleaning materials, and hardware supplies only. Test equipment, such as volt-ohmmeters and oscilloscopes, are detailed in chapter 3, "Electronics Test Equipment." Chapter 4, "Buying Parts," provides a rundown of popular electronic components you will need for project building.

Electronic Tools

Constructing electronic circuit boards requires only a few standard tools. A *soldering iron*, like the one in Fig. 2-1, leads the list. For maximum flexibility, invest in a modular soldering pencil with a 25- to 30-watt heating element. Anything higher could damage electronic components. A 40- to 50-watt element can be used for wiring switches, relays, and power transistors. *Stay away from "instant-on" soldering irons.*

Fig. 2-1. The soldering station. The heat output of this soldering pencil can be controlled by turning the knob.

Although not absolutely required, a soldering iron with variable heat output is a good addition to any electronics bench. In some models, you can vary continuously from low to high; on other models, you can set the heat to low, medium, or high in discrete steps. The heat settings are helpful when adjusting to different sizes of soldering tips.

Supplement your soldering iron with these accessories:

- *Soldering stand*, for keeping the soldering pencil in a safe, upright position. The stand should be made of metal and be constructed so that the top and heating element of the iron does not touch any metal or plastic parts.
- *Soldering tip assortment*. Get one or two small tips for intricate printed-circuit-board work, and get a few larger sizes for routine soldering chores.

- *Solder*. Use resin- or flux-core types only. Acid core and silver solder should *never* be used on electronic components.
- *Sponge*, for cleaning the soldering tip during use. Keep the sponge damp and wipe the tip clean every few uses.
- *Heatsink*, for attaching to sensitive electronic components during soldering. The heatsink draws the excess heat away from the component and helps prevent damage to it.
- *Desoldering vacuum tool*, to soak up molten solder. Use it to get rid of excess solder, remove components, or to redo a wiring job. Buy a good desoldering tool; you'll be using it for many years. The tip receives wear and tear through use and should therefore be replaced as required. Be sure you can obtain replacement parts for your desoldering tool. Note that if you do a lot of desoldering (you salvage components from circuit boards, for example), you might want to invest in a dedicated desoldering station, which includes a soldering iron and permanently attached "solder sucker."
- *Dental picks*, for scraping, cutting, forming, and gouging into the work. You can obtain dental picks at surplus stores and through mail order. Jerryco (see appendix A for address) often sells surplus dental picks for about $3 each.
- *Resin cleaner*. Apply the cleaner after soldering is complete to remove excess resin. The cleaner is available in spray-on cans and brush-on bottles.
- *Solder vise* or "third hand." The vise holds together pieces to be soldered, leaving you free to work the iron and feed the solder. You can purchase the vise with or without a built-in magnifying glass. A typical solder vise is shown in Fig. 2-2. Buy several for elaborate soldering and construction projects.

Wire cutters and strippers

Besides the soldering iron, you'll be using wire cutters and strippers almost exclusively in electronics work. Lightweight cutters and a stripper are shown in Fig. 2-3. These are available at Radio Shack and most other electronics parts stores.The wire cutter is designed to cut flush with a printed circuit board and is useful when you need to get in close. The tool is useful with wire from 30 to 16 gauge. Thicker wire can damage the tool or dull the cutting blades. For thicker wires, use diagonal cutters or "lineman's" pliers.

Wire strippers strip the insulation from your solid or stranded hookup wire. The wire strippers that allow you to "dial in" the gauge of the wire are easier to use. The dial prevents you from cutting or nicking the wire itself, removing just the insulation.

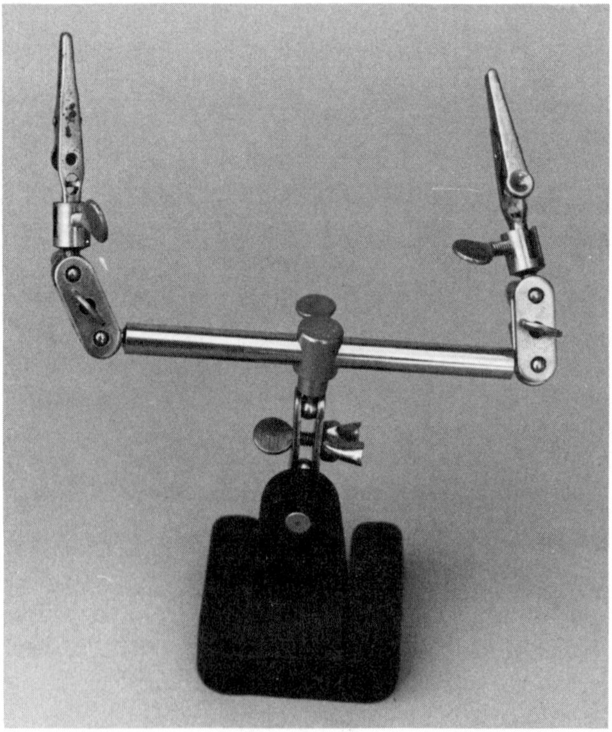

Fig. 2-2. A set of "helping hands," useful in electronics work.

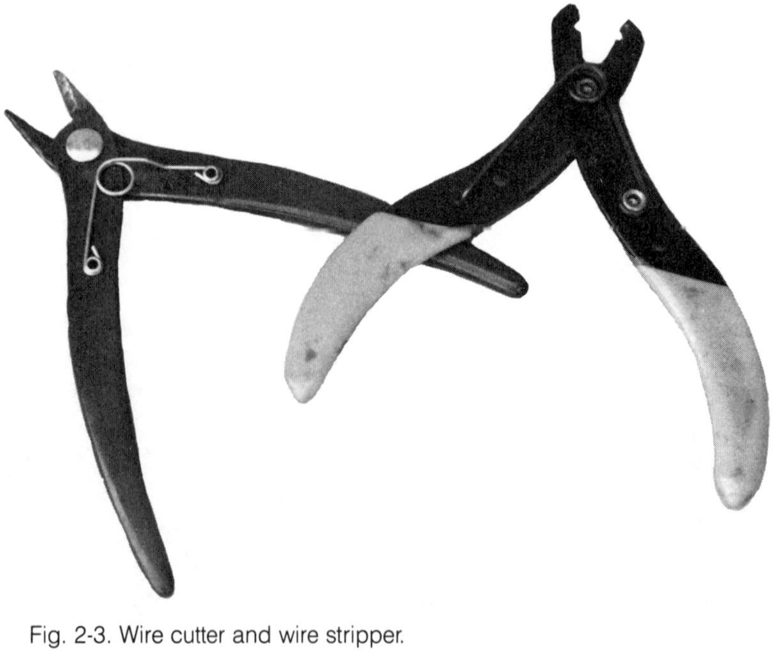

Fig. 2-3. Wire cutter and wire stripper.

Needle-nosed pliers

Pliers help you grip components, bend wires, and hold parts during project assembly. For intricate work, use 5-inch narrow-jaw pliers, like those used in jewelry making (see Fig. 2-4). Larger pliers can be used for general-purpose applications. Be sure that you use the properly sized tool for the job. Using pliers that are too small for the job can ruin the tool. Conversely, using a tool that's too large might damage components or cause undo frustration.

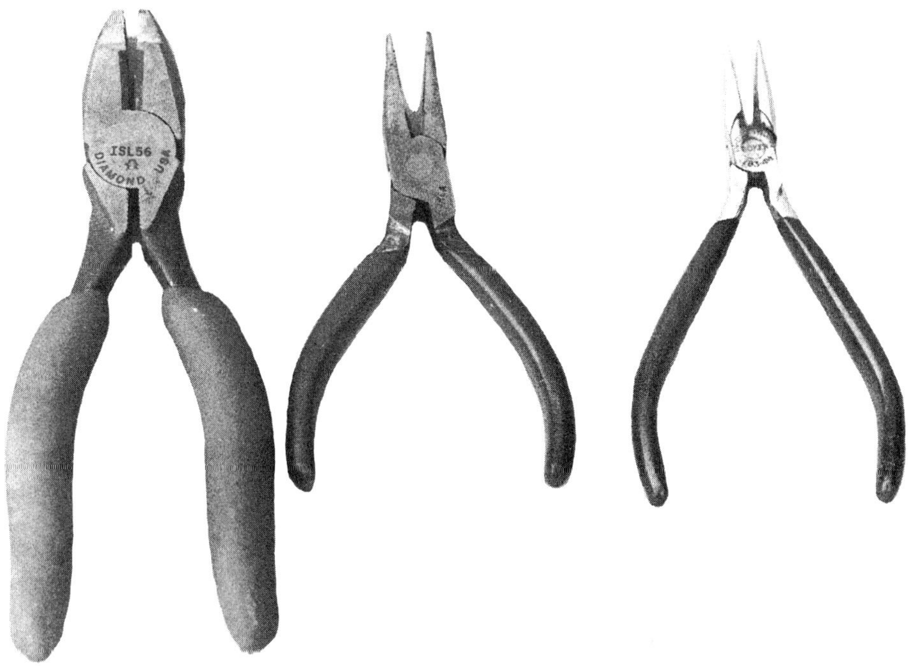

Fig. 2-4. An assortment of pliers.

Adhesives, Glues, and Glue Guns

Many electronics projects require some type of gluing. For example, you might need to secure a small printed circuit board to the inside of a project box. A dab of glue or other adhesive often does the job nicely.

Depending on the application, you can use ordinary household glue, epoxy, cyanoacrylate (''Super'') glue, double-sided foam tape, or hot-melt glue.

- White household glue is available in supermarkets, hardware stores, and home improvements stores. This white stuff comes in small bottles and dries in 10 to 20 minutes (the glue is not ''cured'' for about 12 hours, however). White household glue is susceptible to moisture and is somewhat brittle. It should be used where a firm bond is not absolutely required.

- Epoxy cement comes in two tubes. To use epoxy, mix equal parts of the tubes together, then apply the mixture to the parts you want glued. Most epoxies set up in five minutes and cure completely in one or two hours. Epoxy bonds are strong and resist moisture.
- Cyanoacrylate glue bonds almost anything, almost instantly. Use it with caution, as it can easily bond your fingers together. Use ordinary cyanoacrylate glues when bonding smooth and perfectly matching parts; use the heavier-bodied gap-filling cyanoacrylates if the parts don't mate 100 percent.
- Double-sided foam tape is a quick-and-dirty method of attaching parts. The tape is ideal for securing circuit boards to enclosures or making sure that loosely fitting components remain in place. You can cut the foam tape to almost any size you need, and you can stack layers if you need to fill a large gap. Be sure that the tape and the mating surfaces are dry and free of dirt.
- A hot-melt glue gun is for the person who doesn't like to wait. Slip in a stick of glue, turn the gun on and wait for it to warm up, and you can glue things instantly. The glues are waterproof and gap sealing. The glue is applied at about 250 degrees; which is hot enough to burn skin, but not hot enough to melt solder or damage most electronic components.

Wire and Connectors

In most projects, wire is the stuff that connects all the parts of a circuit together. Hookup wire is generally 20 or 22 gauge and uses a solid conductor, rather than many stranded conductors. Solid wire is preferred for general electronics work because it's easier to solder and unsolder, but it's not as flexible as stranded wire. If the wire will be moved around a lot, use the stranded type to prevent breakage of the conductor. You can purchase solid and stranded hookup wire, in various colors, at most any electronics store.

Connectors let you attach wires to screw terminals or posts. There are literally dozens of connector types, and each has a unique application. Typical connectors for use in electronics projects are the ring tongue and spade. Both attach to a screw terminal.

Quick disconnects let you swiftly engage and disengage an electrical connection. One wire uses a male quick-disconnect terminal; the other uses the female quick-disconnect terminal. The ring tongue and spade connectors, as well as the quick-disconnect terminals, are attached with a crimping tool; soldering is rarely needed. You can also crimp the connectors using a large pair of pliers, but the results aren't as good.

Cleaners and Lubricators

Electronic circuitry must be bright, shiny, and clean to operate properly. After hours of soldering components onto a circuit board, you'll need to clean the board if you want the best electrical contact possible.

The exterior of complete electronics projects, including your test equipment, can be cleaned with a sponge dampened with regular household cleaner, such as Fantastik or 409. Apply the spray to the sponge or cloth, not directly onto the cabinet. Excess can run inside and possibly cause damage.

Circuit boards and the interior of your projects and test gear can be cleaned with an approved electrical cleaner/degreaser. Freon, the stuff used as a coolant in air conditioners, is also a solvent. Unlike most petroleum-based solvents, Freon doesn't melt plastics, and the kind of Freon that you can readily buy is nontoxic and nonflammable (though it does emit chlorinated hydrofluorocarbons, which degrade the ozone layer in the upper atmosphere).

Freon by itself can be used as a basic degreaser and cleaner. You can use it to remove things like grime and dirt in hard-to-reach places. Freon is available at almost any industrial supply house, or if such a business is not nearby, purchase a can of compressed air at a photographic store. Using the can upside down expels the pure Freon propellant.

Freon is often mixed with alcohol to make a more potent cleaner. Use caution with this stuff: it is flammable and toxic. The Freon and alcohol mixture can be used as a general cleaner or degreaser. The Freon/alcohol mixture is available as an all-purpose cleaner/degreaser that you can buy in a spray can. The cleaner is available at Radio Shack and most any electronics supplies store. Figure 2-5 shows a couple of cans that use Freon and alcohol in various mixtures, but they all pretty much do the same thing. The cleaners leave no residue, so you can spray it on and it'll dry with no trace.

The cleaner/degreaser can be used on all the internal components of a project, including the printed circuit board. You can also use the cleaner when maintaining and repairing electrical appliances, like videocassette recorders, compact disc players, and audio amplifiers. Always remember to turn the appliance off and unplug it from the electrical outlet; otherwise, you run the risk of a short circuit.

Some electronic projects require grease and oil. If you plan on servicing radios, VCRs, or similar mechanical devices that use moving parts, keep high-quality lubricating oil and grease at hand.

Just about any light machine oil can be used for components that need oiling. The oil should not have antirust ingredients. Good candidates are 3-in-1 Oil or almost any oil designed for sewing machines. Another good oil is the kind designed for musical instruments. This high-grade oil comes in a handy applica-

Fig. 2-5. An assortment of canned cleaners.

tor bottle. Some bottles have a syringe-type needle for applying the oil in hard-to-reach places. However, the best oil to use is the kind designed for small machined parts. This oil, which is packed in a small bottle with a syringe applicator, has special penetrating lubricants that ordinary oils lack. You can buy this oil at most industrial supply houses and at some camera and electronics stores.

If the oil you have doesn't come with a convenient syringe-type applicator, buy a set of disposal hypodermic needles at the drug store. The needles are usually available behind the counter and can be purchased by adults without a doctor's prescription. The cost is under 50 cents per hypodermic.

To suck the oil up into the hypodermic, open up the can or jar of oil and dip the needle into it. Pull back on the plunger slowly; the oil will sluggishly seep into the hypodermic. To use the syringe, point the end of the needle directly on the spot you want to oil, and push gently on the handle. Apply only a small amount of oil. Exercise care when handling the syringes and always keep the protective caps in place when not in use.

The grease should be a high-quality industrial lubricant, such as Lubriplate, or a non-petroleum-based silicone product. There are a variety of lubricants from which to choose. A light-grade lubricant, such as that used in 35mm cameras, is suitable. You can obtain it from most industrial supply stores and camera repair shops. A small tube goes a long way.

Note that some mechanical components do not require oil or lubrication and in fact can be harmed by them. Motors fall into this category. It's a safe bet that if there is no sign that oil or lubricant has ever been present on a mechanical part, it does not need it!

Miscellaneous Supplies

There are a variety of other cleaning, maintenance, and construction supplies that come in handy when working on electronics:

- Brushes dust out dirt. Any good-quality painter's or artist's brush will do. Stock both a small and a wide brush so you can tackle all jobs.
- Contact cleaner enables you to clean electrical contacts. The cleaner comes in a spray can, but you can apply it by spraying the cleaner onto a brush and whisking the brush against the contacts.
- Cotton swabs soak up excess oil, lubricant, and cleaner. They are available in quantity at any drug store.
- Orange sticks (from a manicure set) and nail files let you scrape junk off circuit boards and electrical contacts.
- The eraser on a pencil goes a long way to rub electrical contacts clean, especially those that have been contaminated by the acid from a leaking battery (however, use care, as rubbing the eraser against the circuit board could create static electricity).
- Modeling putty (for plastic models) can be used to mend cracks and chips on the plastic exterior of your electronic projects.
- A small magnet makes it easier to pick up screws and ferrous metallic objects that have been accidentally dropped into your project.
- A fluid called "Regrip" helps rejuvenate rubber parts. Alcohol should not be used to clean rubber parts, as it can dry them out. Don't use any product (such as "Non-Slip") that leaves a tacky residue.
- A 6× or 8× magnifying glass helps you zoom in close and inspect your handiwork. The magnifying glass is particularly handy when you are looking for solder bridges, cold solder joints, or incomplete joints (these soldering errors are covered more fully in chapter 11, "The Science of Soldering").
- An assortment of batteries and battery holders, in AA, C, D, and 9-volt sizes will ensure that your projects have a ready supply of operating juice. If you find you use batteries a lot, invest in rechargeable nickel-cadmium batteries and a suitable battery recharger.

Construction Tools

Construction tools are the things you use to make printed circuit boards, project enclosures, and housings for your electronic projects. These include hammer, screwdriver, saw, and so forth.

Basic tools

No workshop is complete without the following:

- *Claw hammer*, used for just about anything you can think of.
- *Rubber mallet*, for gently bashing pieces together that resist going together; also for forming sheet metal.
- *Screwdriver assortment*, including various sizes of flathead and Phillips-head screwdrivers. A few long-blade screwdrivers are handy to have, as well as a ratchet driver. Get a screwdriver magnetizer/demagnetizer; it lets you magnetize the blade so it attracts and holds screws for easier assembly. For precision work, invest in a set of jeweler's tools with both flathead and Phillips blades.
- *Hacksaw*, to cut anything. The hacksaw is the staple of the electronics hobbyist. Get an assortment of blades. Coarse-tooth blades are good for wood and PVC pipe plastic; fine-tooth blades are good for copper, aluminum, and light-gauge steel.
- *Miter box*, to cut straight lines. Buy a good miter box and attach it to your worktable (avoid wood miter boxes; they don't last). You can also use the box to cut stock at near-perfect 45 degree angles, helpful when building optical benches.
- *Wrenches*, all types. Adjustable wrenches are helpful additions to the shop, but careless use can strip nuts. The same goes for long-nosed pliers, useful for getting at hard-to-reach places. A pair or two of vise grips (indispensable in my workshop) helps you hold pieces for cutting and sanding. A set of nut drivers make it easy to attach nuts to bolts.
- *Measuring tape*. A six- or eight-foot steel measuring tape is a good choice. Also get a cloth tape at a fabric store for measuring flexible things.
- *Square*, for making sure that pieces you cut and assemble from wood, plastic, and metal are square.
- *File assortment*, to smooth the rough edges of cut wood, metal, and plastic (particularly important when working with metal because the sharp unfinished edges can cut you).
- *Drill motor*. Get one that has a variable speed control (reversing is nice, but not absolutely necessary). If the drill you have isn't variable speed, buy a variable speed control for it. You need to slow the drill when work-

ing with metal and plastic. A fast drill motor is good for wood only. The size of the chuck is not important because most of the drill bits you'll be using will fit a standard 1/4-inch chuck. For light-duty, intricate work, you can use a hand-operated drill.

- *Drill bit assortment*. Use good, sharp ones *only*. If yours are dull, have them sharpened (or do it yourself with a drill-bit sharpening device), or buy a new set. *Never* use dull drill bits.
- *Vise*, for holding parts while you work. A large vise isn't required, but you should get one that's big enough to handle the size of pieces you'll be working with.
- *Clamps* hold materials for you and give you an extra hand.
- *Clear safety goggles*. Wear them when hammering, cutting, drilling, and any other time flying debris could get in your eyes. Be sure you use the goggles.

If you plan on using wood in any of your projects (a wooden base for an old-time crystal radio, for instance), you might want to consider adding rasps, wood files, coping saws, and other woodworking tools to your toolbox. Working with plastic requires a few extras, as well, including a burnishing wheel to smooth the edges of the cut plastic (a flame from a cigarette lighter also works but is harder to control), a strip heater for bending, and special plastic drill bits. These bits have modified tips that aren't as likely to rip through the plastic material. You can get bits for glass as well. You can cut or score small plastic parts using a sharp razor knife or razor saw, available at hobby stores.

Optional tools

There are a number of tools you can use to make your time in the electronics shop more productive and less time consuming. A *drill press* helps you drill better holes because you have more control over the angle and depth of each hole. Use a drill press vise to hold the pieces; never use your hands. The drill press is particularly handy when constructing your own printed circuit boards. Outfitted with a small #58 twist drill, you can quickly and efficiently drill the component mounting holes in the circuit board.

A *table saw* or *circular saw* makes cutting through large pieces of wood and plastic easier. Use a guidefence, or fashion one out of wood and clamps, to ensure a straight cut. Be sure to use a fine-tooth saw blade if cutting through plastic. Using a saw designed for general wood cutting will cause the plastic to shatter.

A motorized *hobby tool* is much like a handheld router. The bit spins very fast (25,000 rpm and up), and you can attach a variety of wood, plastic, and metal working bits to it. The better hobby tools, such as those made by Dremel and

Weller, have adjustable speed controls. Use the right bit for the job. For example, don't use a wood rasp bit with metal or plastic because the flutes of the rasp will too easily fill with metal and plastic debris.

A *nibbling tool* is a fairly inexpensive accessory (under $20) that lets you "nibble" small chunks from metal and plastic pieces. The maximum thickness depends on the bite of the tool, but it's generally about 1/16 or 1/8 inch. Use the tool to cut channels, enlarge holes, and so forth. A *tap and die set* lets you thread holes and shafts to accept standard size nuts and bolts. Buy a good set. A cheap assortment of taps and dies is more trouble than its worth.

A *thread-size gauge* made of stainless steel might seem expensive ($10 to $12), but it helps you determine the size of any standard SAE or metric bolt. It's a great accessory for use with taps and dies. Most gauges can be used when chopping threads off bolts with a hacksaw, providing a cleaner cut.

A *brazing tool* or *small welder* lets you spot weld two metal pieces together. These tools are designed for small pieces only. They don't provide enough heat to adequately weld pieces larger than a few inches in size. Be sure that extra fuel and/or oxygen cylinders or pellets are readily available for the brazer or welder you buy. There's nothing worse than spending $30 to $40 for a home welding set, only to discover that supplies are not available for it. Be sure to read the instructions that accompany the welder and observe all precautions.

Hardware Supplies

Many of the projects you'll build will consist of standard electronic parts as well as frames and mechanical subsystems. For example, if you are building a robot, you'll need to construct frames, battery holders, appendages, and so forth. If you're building a modernized Tesla coil, you'll need to construct the base, the primary and secondary coils, the spark gap, and most likely, the high-voltage capacity.

You can obtain the materials for these subsystems at most hardware or home improvement stores. Here are some common items you'll want to have around your shop.

Nuts and bolts

Number 8 and 10 nuts and panhead stove bolts (8/32 and 10/24, respectively) are good for all-around construction. Get a variety of bolts in 1/2-, 3/4-, 1-, 1 1/4-, and 1 1/2-inch lengths. You may also want to get some 2-inch and 3-inch long bolts for special applications. For smaller applications, use 4-40 or 6-32 hardware. Very small nuts and bolts can be found at hobby stores specializing in radio controlled model airplanes.

Motor shafts and other heavy duty applications require 1/4-inch 20 or 5/16-inch hardware. Panhead stove bolts are the best choice; you don't need hexhead carriage bolts unless you have a specific requirement for them.

Washers

While you're at the store, stock up on flat washers, fender washers (large washers with small holes), tooth lockwashers and split lockwashers. Get an assortment for the various sizes of nuts and bolts. Split lockwashers are good for heavy-duty applications because they provide more compression locking power. You usually use them with bolt sizes of 1/4 inch and above.

All-thread rod

All-thread is 2- to 3-foot lengths of threaded rod stock. It comes in standard thread sizes and pitches. All-thread is good for shafts and custom-made fasteners. Get one of each in 8/32, 10/24, and 1/4-inch 20 threads to start.

Special nuts

Coupling nuts are just like regular nuts but have been stretched out. They are designed to couple two bolts or pieces of all-thread together, end to end. Locking nuts have a piece of nylon built into them that provides a locking bite when threaded onto a bolt. Locking nuts are preferred over using two nuts tightened together.

Rivets

You don't have to be Rosy the Riveter to use (and appreciate) rivet fasteners. Most hardware stores offer a wide variety of steel and aluminum rivets in various diameters and lengths. The rivets are the "blind" type, meaning you do not need access to the front and back ends of the rivet in order to secure it into position. To insert a blind rivet, place the "stalk" of the rivet in a special tool (see Fig. 2-6), and the head of the rivet into a hole in the pieces you want to fasten. Squeeze the tool closed; the head of the rivet mushrooms out and clamps the pieces together, as shown in Fig. 2-7. Continue squeezing and the stalk of the rivet breaks off.

While you need to invest in a special riveting tool (cost is about $10 to $20, depending on quality), the main advantages of using rivets is that they weigh less and assembly time is faster, especially if you are fastening two pieces where it would otherwise be difficult to use a separate bolt and nut. The main disadvantage to rivets is that they are permanent. You must drill or break the rivet out of the hole if you want to disassemble the pieces.

Extruded aluminum

For most of your projects that require a metal framework, you can take advantage of a rather common hardware item: extruded aluminum stock. This stuff,

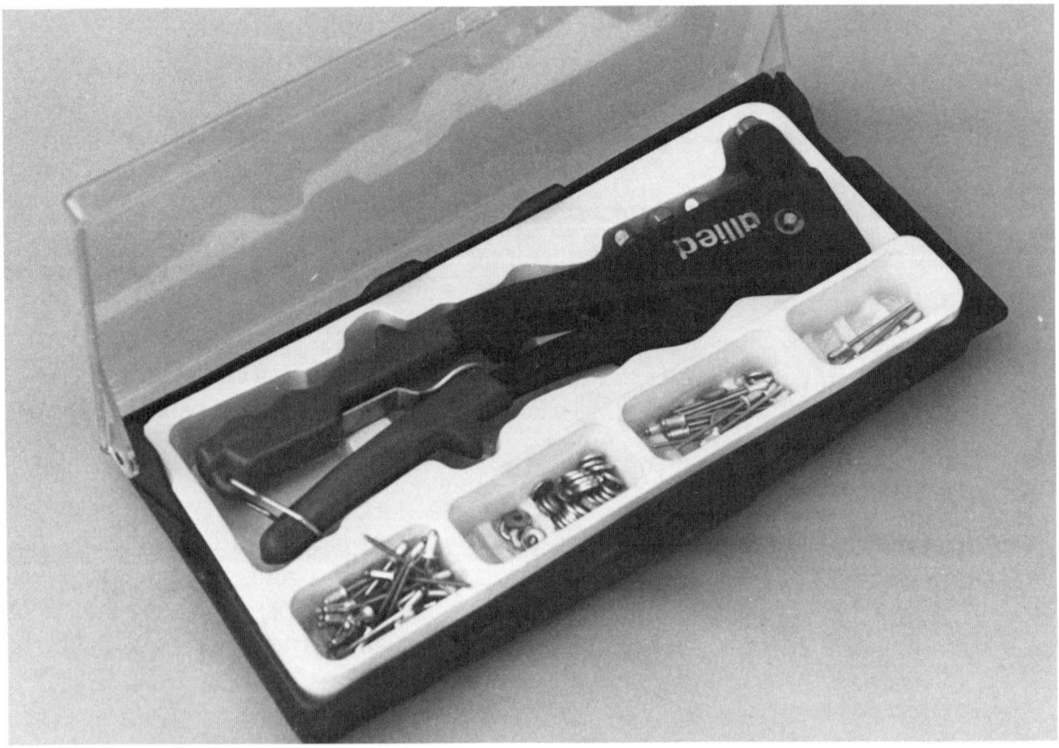

Fig. 2-6. Rivet tool.

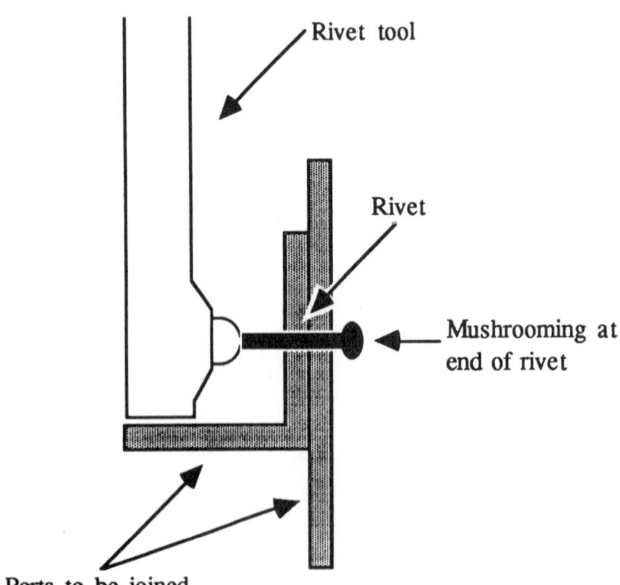

Fig. 2-7. How blind (or "pop") rivets work.

which is designed for such things as building bathtub enclosures, picture frames, and other handyman applications, comes in various sizes, thicknesses, and configurations. The length is usually 6, 8, 10, or 12 feet, but if you need less, most hardware stores will cut to order (you save when you buy it in full lengths). The stock is available in plain (dull silver) anodized aluminum and gold anodized aluminum. Get the plain stuff: it's 10 to 25 percent cheaper.

Two particularly handy stocks are 41/64- by 1/2- by 1/16-inch channel and 57/64- by 9/16- by 1/16-inch channel (some stores sell similar material with slightly different dimensions). I use these extensively for parts for robots, lasers, high-voltage devices, and many other projects.

Angle stock measuring 1 × 1 × 1/16 inches is another often-used item, usually employed for attaching cross bars and other structural components. No matter what size you eventually settle on for your own designs, keep several feet of angle stock handy at all times. You can always use it.

If extruded aluminum is not available, another approach is to use shelving standards, the bar-like channel stock used for wall shelving. It's most often available in steel, but some hardware stores carry it in aluminum (silver, gold, and black anodized).

The biggest problem with using shelving standards is that the slots can cause problems when drilling holes for hardware. The drill bits can slip into the slots, causing the hole to be off center. Some standards have an extra lip on the inside of the channel that can interfere with some of the hardware you might use to join the pieces together.

Angle brackets

You need a good assortment of 3/8-inch and 1/2-inch galvanized iron brackets to join the extruded stock or shelving standards together. Use 1 1/2-inch by 3/8-inch flat corner irons when joining pieces cut at 45-degree angles to make a frame. The 1-inch by 3/8-inch and 1 1/2-inch by 3/8-inch corner angle irons are helpful when attaching the stock to baseplates and when securing various components.

3

Electronics test equipment

NO ELECTRONICS BENCH IS COMPLETE WITHOUT TEST EQUIPMENT. IF YOU'RE just starting out, your test equipment might consist of a volt-ohmmeter and little else. As you become more committed to your electronics hobby, you can add a logic probe, logic pulser, oscilloscope, frequency counter, signal generator, and other odds and ends that help you design, build, and troubleshoot circuits.

This chapter provides an overview of popular electronic bench test equipment. You'll find details on several types of test equipment, including the all-important volt-ohmmeter. You'll also learn about oscilloscopes and how they are used, plus details on special-purpose bench tools like sweep generators, signals injectors, and logic probes.

Volt-ohmmeter

A *volt-ohmmeter* is used to test voltage levels and the impedance of circuits. This moderately priced electronic tool is a basic requirement for all electronics work. If you don't already own a volt-ohmmeter, you should seriously consider buying one. The cost is rather minimal considering the usefulness of the device.

There are many volt-ohmmeters (or VOMs) on the market today. For basic electronics experiments, you don't want a cheap model and you don't need an expensive one. A meter of intermediate quality is sufficient and does the job admirably. The price for such a meter is between $30 and $75 (it tends to be on the low side of this range). Meters are available at Radio Shack and most electronics outlets. Shop around and compare features and prices.

Digital or analog

There are two general types of VOMs available today: digital and analog. The difference is not that one meter is used on digital circuits and the other on analog circuits. Rather, digital meters employ a numeric display like a digital clock or watch. Analog VOMs use the older-fashioned—but still useful—mechanical movement with a needle that points to a set of graduated scales.

Digital VOMs used to cost a great deal more than the analog variety, but the price difference has evened out recently. Digital VOMs, such as the one shown in Fig. 3-1, are quickly becoming the standard; in fact, it's hard to find a decent analog meter anymore. Digital VOMs also tend to be lighter and more compact than their analog cousins. The VOM in Fig. 3-2 measures just $2 \times 4\frac{1}{4} \times \frac{3}{8}$ inches (without pocket carrying case), yet does just about everything you need a volt-ohmmeter for.

Analog VOMs are traditionally harder to use because you must select the type and range of voltage you are testing, find the proper scale on the meter face, and then estimate the voltage as the needle swings into action. Digital VOMs, on the other hand, display the voltage in clear numerals and with a greater precision than most analog meters. Because of their increased popularity and ease of use, this chapter concentrates on digital VOMs exclusively.

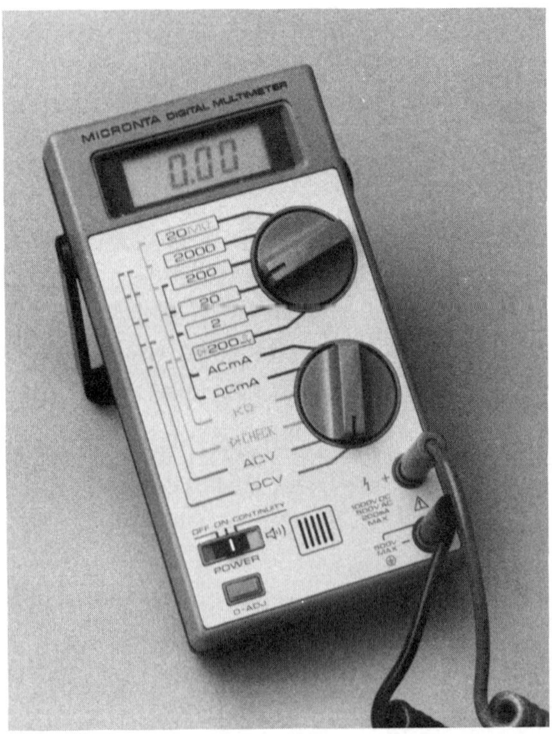

Fig. 3-1. A typical digital VOM.

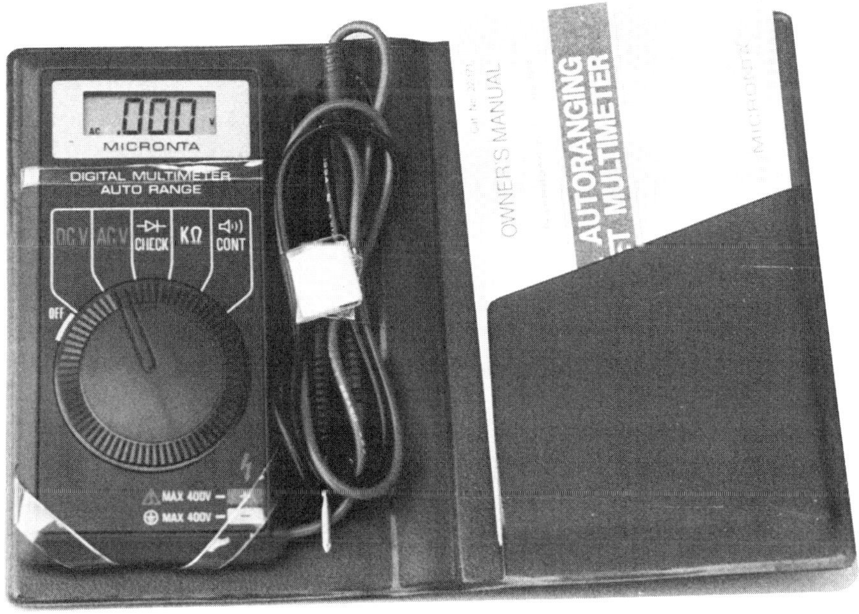

Fig. 3-2. A pocket VOM.

Automatic ranging

As with analog meters, some digital meters require you to select the range before it can make an accurate measurement. For example, if you are measuring the voltage of a 9-volt transistor battery, set the range to the setting closest to and above 9 volts (with most meters it is the 20- or 50-volt range). Autoranging meters don't require you to do this, so they are inherently easier to use. When you want to measure voltage, set the meter to volts (either ac or dc) and take the measurement. The meter displays the results in the readout panel.

While autoranging is generally preferable to manual ranging, it does have one drawback: slower reaction time. With autoranging, the meter takes a little while longer to display the result in the readout. This is because the meter is ready to accept the entire range of voltages or resistances it can accommodate, rather than a more narrow band. However, even though autoranging meters tend to be slower, the extra wait is not always noticeable—often a tenth of a second or better.

Accuracy

Little of the work you'll do with general electronics require a meter that's super accurate. A VOM with average accuracy is more than enough. The accuracy of a meter is the minimum amount of error that can occur when making a specific measurement. For example, the meter might be accurate to 2000 volts ± 0.8 percent. The 0.8 of 1 percent error at the kinds of voltages used in most dc-operated circuits—typically 5 to 12 volts dc—is only 0.096 volts.

Digital meters have another kind of accuracy. The number of digits in the display determines the maximum resolution of the measurements. Most digital meters have 3½ digits, so it can display a value as small as 0.0001 (the half digit is a "1" on the left side of the display). Anything less than that is not accurately represented, but there's little cause for accuracy higher than this when working in hobby electronics.

Functions

Digital VOMs vary greatly in the number and type of functions they provide. At the very least, all standard VOMs let you measure ac volts, dc volts, milliamps, and ohms. Some also test capacitance, opens, or shorts in discrete components like diodes and transistors. A multifunction meter is shown in Fig. 3-3. It includes a component tester for capacitors, diodes, and transistors.

Fig. 3-3. A multifunction VOM. In addition to the basic VOM features, this one tests transistors and capacitors.

For most purposes, these additional functions are not necessary, and you need not spend the extra money on a meter that includes them. To make effective measurements, you need to take diodes and transistors out of the circuit to accurately test them. The additional testing features of multifunction meters can come in handy, however, when testing components during the design or construction stage of a project. For example, if you need a capacitor that is *exactly* 0.1 μF, you can fish through your pile of capacitors and test all those marked for 0.1 μF for the greatest degree of tolerance.

The maximum range of the meters when measuring volts, milliamps, and resistance also varies. For most applications, the following maximum ratings are more than adequate:

dc volts	1,000 volts
ac volts	500 volts
dc current	200 milliamps
resistance	2 megohms

If your meter can also test capacitors, a range of 20 μF to 20 pF is desirable.

Meter supplies

Most meters come with a pair of test leads—one black and one red—each equipped with a needle-like metal probe. The quality of the test leads is usually minimal, so you might want to purchase a better set. The kind with coiled leads are handy. They stretch out to several feet yet recoil to a manageable length when not in use.

Standard leads are fine for most routine testing, but some measurements require the use of a clip lead. These have a spring-loaded clip on the end; you can clip the lead in place so your hands are free to do other things. The clips are insulated to prevent short circuits, and you can get clips that attach onto regular test leads.

Meter safety and use

Most applications of the meter involve testing low voltages and resistance, both of which are relatively harmless to humans. Sometimes, however, you need to test high voltages—like the input to a power supply—and careless use of the meter can cause serious bodily harm. Even when you're not actively testing high voltage, it might be exposed when you remove the protective cover of the circuit.

If the circuit is plugged in while the cover is off (which it will be if you're testing for proper operation) and you're not measuring voltages at the power supply, cover the power supply terminals, if exposed, with a piece of cardboard or insulating plastic.

Proper procedures for meter use involves setting the meter beside the unit under test, making sure it is close enough so that the leads reach the test points inside the deck. Plug in the leads and test the meter operation by first selecting the resistance function setting (use the smallest scale if the meter is not autoranging). Touch the leads together. The meter should read 0 ohms.

If the meter does not respond, check the leads and internal battery and try again. If the display does not read 0 ohms, double-check the range and function settings and adjust the meter to read 0 ohms (not all digital meters have a zero adjust, but most analog meters do).

Once the meter has checked out, select the desired function and range and apply the leads to the circuit under test. Usually, the black lead is connected to ground, and the red lead is connected to the various test points in your circuit. Figure 3-4 shows a test of a VCR power supply.

Never blindly poke at a circuit in an attempt to get some kind of reading. Apply the test leads only to those portions of the circuit you are familiar with. If you didn't design the circuit, keep the schematic handy (if one is available). Refer to the schematic for the location of test points.

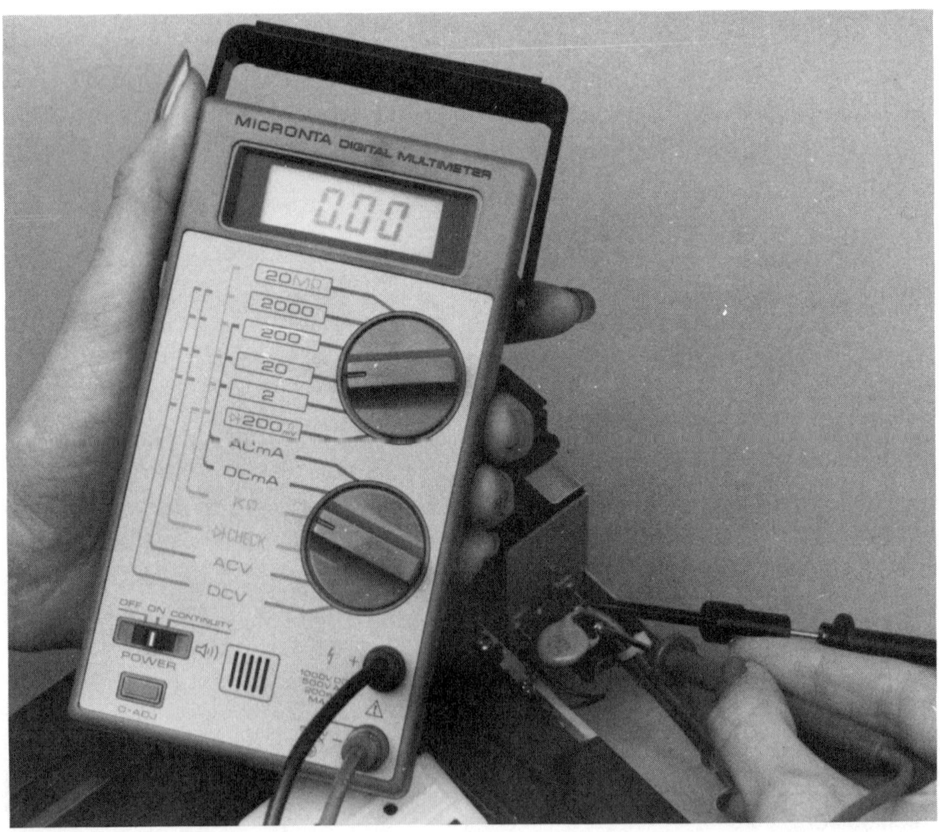

Fig. 3-4. How to use a meter.

One safe way to use the meter is to attach a clip on the ground lead and connect it to the chassis or circuit ground. Use one hand to apply the red lead to the various test points; stick the other hand safely in your pocket. With one hand "out of commission," you are less likely to receive a bad shock because you weren't watching what you were doing.

Important: Always avoid the situation where you are touching a ground contact and a live ac circuit at the same time. The electricity will flow right through you, including your heart, and it could be the last thing you ever do.

Logic Probe

Meters are typically used for measuring analog signals. Logic probes test for the presence or absence of the low-voltage dc signals that represent digital data. These 0s and 1s are usually electrically defined as 0 and 5 volts, respectively (although the actual voltages of 0 and 1 bits depends entirely on the circuit). You can use a meter to test a logic circuit, but the results aren't always predictable. Further, many logic circuits change states quickly (pulse) and meters cannot track the voltage switches fast enough.

Logic probes, such as the model in Fig. 3-5, are designed to give a visual and (sometimes) aural signal of the logic state of a particular circuit line. One LED on the probe lights up if the logic is 0 (or low), and another LED lights up if the logic is 1 (or high). Most probes have a built-in buzzer, which has a different tone for the two logic levels. That way, you don't need to keep glancing at the probe to see the logic level.

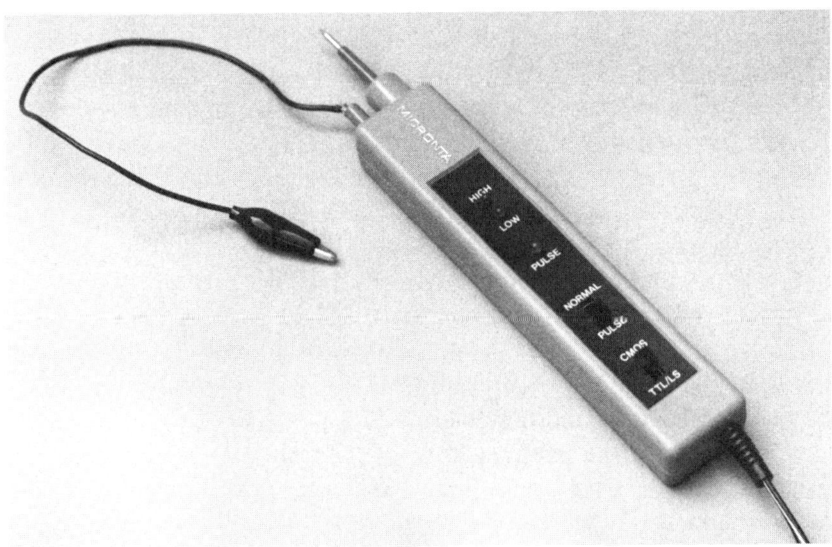

Fig. 3-5. A logic probe.

A third LED or tone might indicate a pulsing signal. A good logic probe can detect that a circuit line is pulsing at speeds of up to 10 MHz, which is more than fast enough for most hobby electronics applications. The minimum detectable pulse width (the time the pulse remains at one level) is 50 nanoseconds, again more than sufficient for routine design and troubleshooting work.

Although logic probes might sound complex, they are really simple devices, and their cost reflects this. You can buy a reasonably good logic probe for under $20. Most probes are not battery operated; rather, they obtain operating voltage from the circuit under test.

The usefulness of a logic probe depends on the type of circuits you build and test. If you seldom work with digital circuits (which is rare today), a logic probe will have limited application for you. The more you experiment with digital logic, the more beneficial a logic probe can be.

Using a logic probe

The same safety precautions apply when using a logic probe as they do when using a meter. When the cover of the project is removed, potentially dangerous high voltages might be exposed. If you are working close to these voltages, cover them to prevent accidental shock. Logic probes cannot operate with voltages exceeding about 15 volts dc, so if you are unsure of the voltage level of a particular circuit, test it with a meter first to be sure it is safe.

Successful use of the logic probe really requires you to have the circuit schematic to refer to. It's nearly impossible to blindly use the logic probe on a circuit without knowing what you are testing. A single circuit board might contain components for both digital and analog signal processing, and you must know exactly what each component does and how it is used. And, since the probe receives its power from the circuit under test, you need to know where to pick off suitable power. You can easily damage the probe—and possibly the circuit under test—if you connect the power leads incorrectly.

To use the probe, connect the probe's power leads to a voltage source on the board, clip the black ground wire to circuit ground, and touch the tip of the probe against a pin of an integrated circuit or the lead of some other component. Figure 3-6 shows a probe testing the logic level at an IC pin.

Please note: many circuits, particularly those that employ operational amplifiers, often use negative voltages with respect to ground. To the logic probe, there is little difference between connecting the power leads to a positive and ground rail or connecting the power leads to a negative and ground rail. However, avoid connecting the power leads to a positive *and* negative rail, because the voltage differential might exceed the maximum supply voltage of the probe. For instance, connecting the lead between the +9 and −9 rails will feed 18 volts to the probe, which is higher than its rated operating voltage.

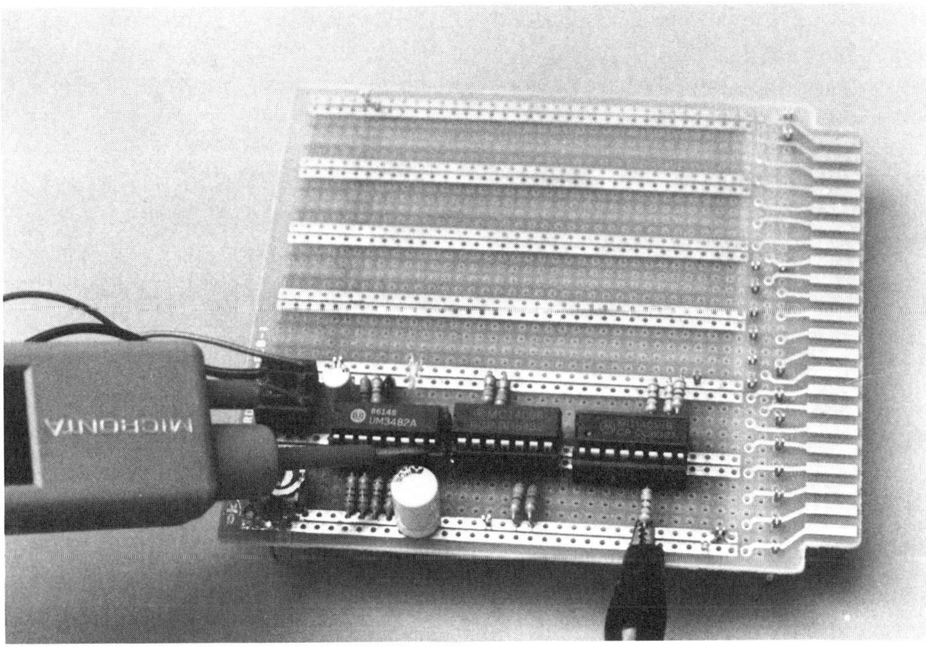

Fig. 3-6. Using the logic probe.

Logic Pulser

A handy troubleshooting accessory when working with digital circuits is the *logic pulser*. This device puts out a timed pulse, letting you see the effect of the pulse on a digital circuit. Normally, you'd use the pulser with a logic probe or an oscilloscope (discussed below). The pulser is switchable between one pulse and continuous pulsing.

Most pulsers obtain their power from the circuit under test. It's important that you remember this. With digital circuits, it's generally a bad idea to present an input signal to a device that's greater than the supply voltage for that device. In other words, if a chip is powered by 5 volts and you give it a 12-volt pulse, you'll probably ruin the chip. Some circuits work with split (+, −, and ground) power supplies, so be sure you connect the leads of the pulser to the correct power points.

Also be sure that you do not pulse a line that has an output but no input. Some integrated circuits are sensitive to unloaded pulses at their output stages, and improper application of the pulse can destroy the chip.

Oscilloscopes

An *oscilloscope* is a pricey tool—good ones start at about $400—and only a small number of electronic hobbyists own one. For really serious work, however, an oscilloscope is an invaluable tool, one that will save you hours of time and frustration. Things you can do with a scope include some of the things you can do with other test equipment, but oscilloscopes do it all in one box and generally with greater precision. Among the many applications of an oscilloscope, you can:

- Test dc or ac voltage levels.
- Analyze the waveforms of digital and analog circuits.
- Determine the operating frequency of digital, analog, and RF circuits.
- Test logic levels.
- Visually check the timing of a circuit to see if things are happening in the correct order and at the prescribed time intervals.

A basic, no-nonsense oscilloscope is shown in Fig. 3-7. This entry-level oscilloscope is a dual-trace (two channel) model with a 20 MHz bandwidth. The two channels let you monitor two signals at once so you can easily view the input signal and output signal at the same time.

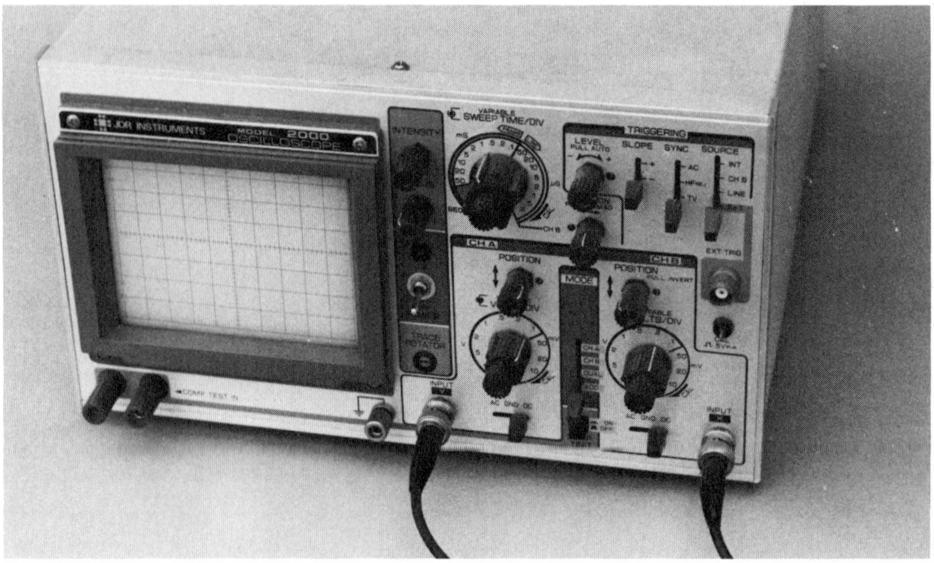

Fig. 3-7. A typical 20 MHz oscilloscope. This is a good, basic model, readily available from a variety of sources.

The 20 MHz bandwidth is a guide to the signals the scope can track—from 0 Hz to 20 MHz. The upper range is the more important (0 Hz simply means dc voltage) and indicates the fastest circuits the scope can be used with. Circuits that operate in excess of 20 MHz are beyond the range of the scope, and cannot be tested. Realistically, few hobby circuits operate at speeds faster than 5 MHz, let alone 20 MHz, so there is plenty of slack built in.

Scopes are not particularly easy to use; they have lots of dials and controls that set operation. For this reason, their use is not discussed here; rather, read the instruction manual that comes with your particular scope for details on how to operate it.

Figure 3-8 details the controls on a basic oscilloscope and summarizes their function. Thoroughly familiarize yourself with the operation of your oscilloscope before using it. Knowing how to set the time-per-division knob is as important as knowing how to turn the scope on. As usual, exercise caution when using the scope with or near high voltages. Generally, the same safety precautions outlined above for volt-ohmmeters also applies to the use of oscilloscopes.

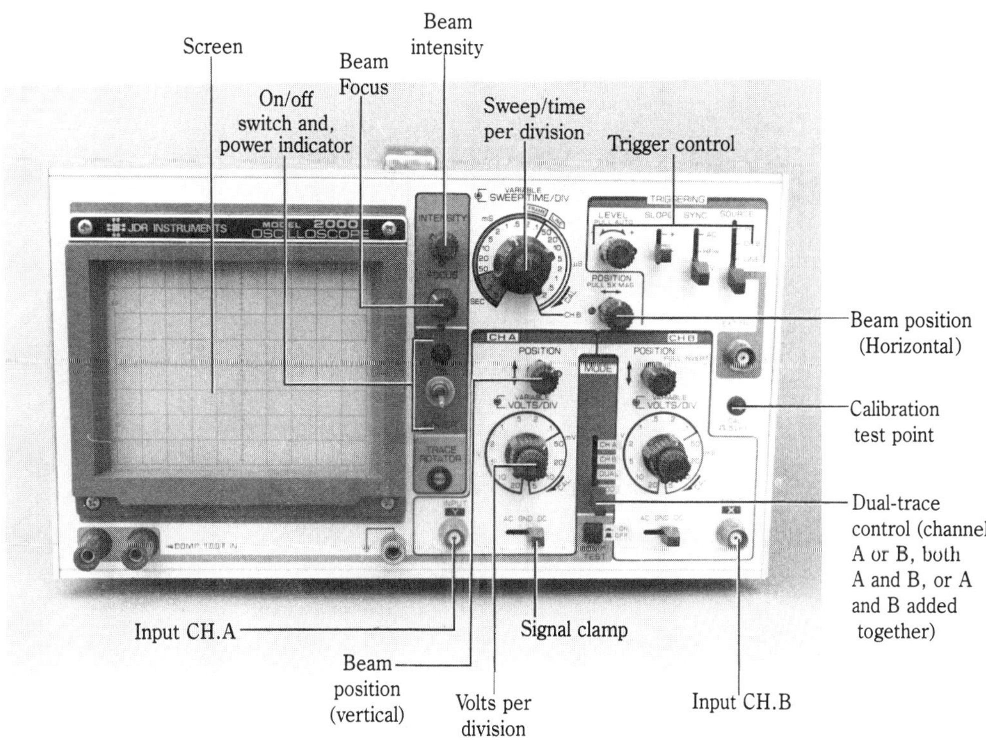

Fig. 3-8. The controls of the oscilloscope. Get to know them well.

Oscilloscope bandwidth and resolution

One of the most important specifications of an oscilloscope is its bandwidth. If 20 MHz is too low for your application, you should invest in a more expensive oscilloscope with a bandwidth of 35, 60, or even 100 MHz. Price goes up considerably as the bandwidth increases.

The resolution of the scope reveals its sensitivity and accuracy. On an oscilloscope, the x-axis (horizontal) displays time, and the y-axis (vertical) displays voltage. The sweep time indicates the x-axis resolution, generally 0.5 microseconds or faster. The sweep time is adjustable so you can test signal events that occur over a longer time period, usually as long as a half a second to a second. Note that signal events faster than 0.5 microseconds can be displayed on the screen, but the signal might appear as a fleeting glitch or voltage spike.

The sensitivity indicates the y-axis resolution. The low-voltage sensitivity of most average-priced scopes is about 5 mV to 5 volts. Turn the dial to set the sensitivity you want. When setting the dial to 5 mV, for example, each tick mark on the face of the scope tube represents a difference of 5 mV, as shown in Fig. 3-9. Voltage levels lower than 5 mV might appear, but they cannot be accurately measured. Most scopes show very low level voltages (microvolt range) as a slight ripple.

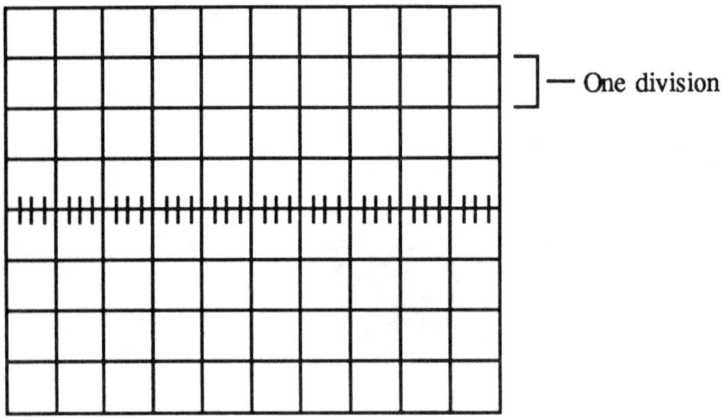

Fig. 3-9. How the screen of the oscilloscope is marked with graduations.

Enhanced features

Over the years, oscilloscopes have improved dramatically, with many added features and capabilities. Among the most useful features are:

- *Delayed sweep* is helpful when analyzing a small portion of a long, complex signal. Ideal for use with television signals, as you can zoom into just a portion of the signal and examine it.
- *Digital storage* records signals in computerized memory for later recall. Once in the memory, you can expand the signal and analyze specific portions; again, helpful in television work. Digital storage also lets you compare signals, even if you take the measurements at different times.
- *Selectable triggering* lets you choose how the scope triggers on the input signal. When checking dc signals, no triggering is necessary, but ac and digital signals require that you select a specific part of the signal so that

the scope can properly display the waveform. At the minimum, a scope provides automatic triggering that locks onto most stable ac and digital signals. This is internal triggering that detects the rise or fall of the input signal. If you're doing lots of TV work, get a scope that lets you trigger on horizontal and/or vertical sync video pulses. External triggering permits the oscilloscope to use an auxiliary signal for locking onto the primary signal. On most dual-trace scopes, you can select channel A or B as the trigger signal.

Use good scope probes

The probes used with oscilloscopes are not just wires with clips on their ends. To be effective, the better scope probes use low-capacitance/low-resistance shielded wire and a capacitive-compensated tip, both of which ensure better accuracy. Consider the waveforms in Fig. 3-10. The waveform on the top is typical of the response provided by a simple wire probe. Capacitive effects distort the signal and mask its true form. The waveform on the bottom illustrates the corrected waveform as it really appears.

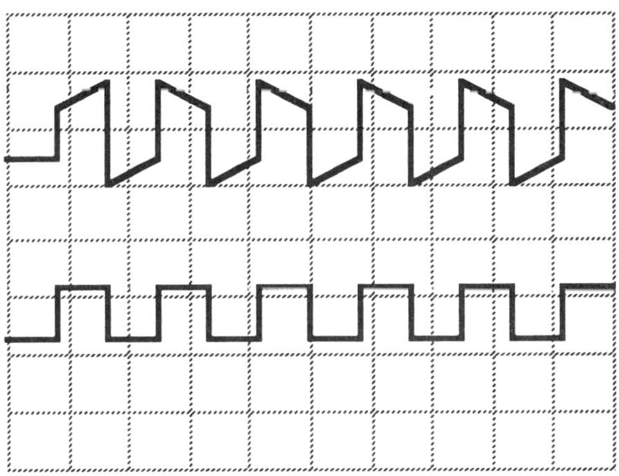

Fig. 3-10. Examples of good and bad capacitive-loaded waveforms.

Most scope probes are passive, meaning that they employ a simple circuit of capacitors and resistors to compensate for the effects of capacitive and resistive loading. Figure 3-11 shows the components used in the average scope probe. Many passive probes are switchable between 1× and 10×. At the 1× setting, the probe passes the signal without attenuation (weakening). At the 10× setting, the probe reduces the signal strength by 10 times. This allows you to test a signal that might otherwise overload the scope's circuits. Note that oscilloscopes and scope probes are seldom rated for more than 600 volts peak to peak, so they cannot be used on very high voltage circuits.

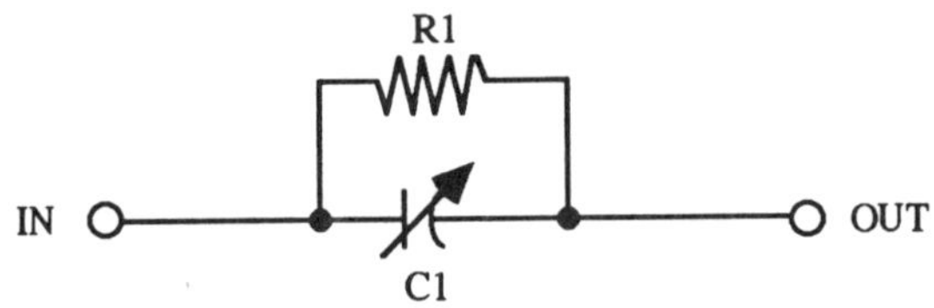

Fig. 3-11. A basic scope probe consists of a resistor and capacitor.

Active probes use operational amplifiers or other powered circuitry to correct for the affects of capacitive and resistive loading and to vary the attenuation of the signal. Table 3-1 shows the typical specifications of passive and active oscilloscope probes.

Table 3-1. Typical Oscilloscope Probe Specs

Probe Type	Frequency Range	Resistive Load	Capacitive Load
Passive 1×	dc – 5 MHz	1 megohm	30 pF
Passive 10×	dc – 50 MHz	10 megohms	5 pF
Active	dc – 500 MHz	10 megohms	2 pF

Frequency Counter

A *frequency counter* (or frequency meter) tests the operating frequency of a circuit and, like the oscilloscope, is not an absolute requirement for beginner-level electronics work. Most models, like the one shown in Fig. 3-12, can be used on digital, analog, and RF circuits for a variety of testing chores. You need only a basic frequency counter—a $100 to $200 investment. You can save some money by building a frequency counter kit. Plans for a basic, no-frills frequency counter is provided in chapter 20.

Frequency counters have an upward operating limit, but it's generally well within the region applicable to hobby electronics. A frequency counter with a maximum range of up to 50 MHz is generally more than enough. Higher priced meters come with or have as options a "prescaler," a device that extends the useful operating frequency to 500 MHz and above.

Miscellaneous Test Gear

The well-equipped electronics lab also includes the following test equipment. These items are optional and should be purchased only when the need for them arises.

Variable power supply The variable power supply provides a well-regulated voltage output, generally from 0 to 20 volts. The model shown in Fig. 3-13 offers a variable output from about 2 to 20 volts, as well as preset outputs of –5,

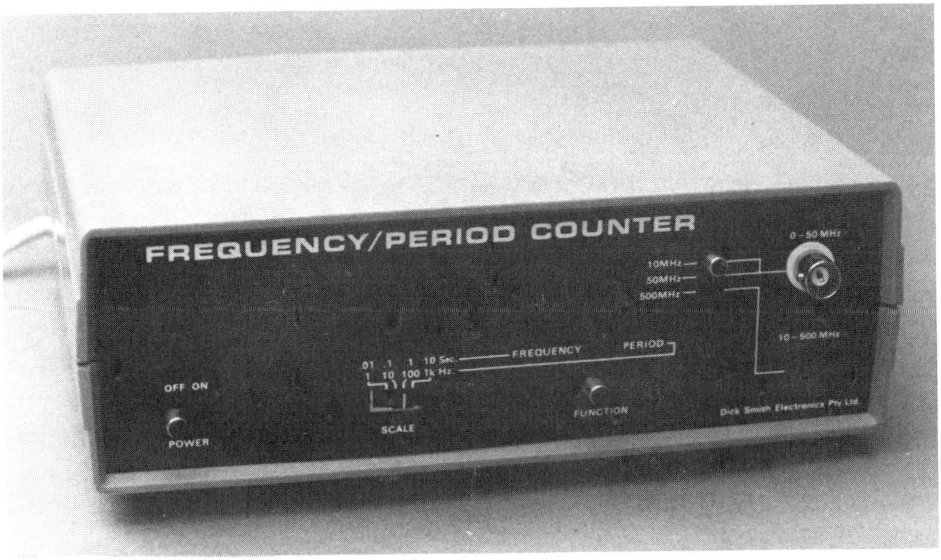

Fig. 3-12. A frequency counter.

Fig. 3-13. A variable power supply. This model also delivers 5 and 12 volts.

+5, and +12 volts. Your power supply should be able to provide at least 1 amp per circuit, preferably more. Several bench power supply circuits are provided in chapter 20.

Function generator A function generator creates nearly pure signal waveforms for testing and calibration purposes. Most function generators develop three kinds of waveforms: sine, triangle, and square. The frequency of the waveforms can be adjusted from a low of 1 Hz to a high of 20 to 50 kHz.

Accurate timing of the waveform is possible only with a frequency counter. If you have a separate frequency counter, you can use it to fine-tune the output of the function generator. Some function generators come with a frequency counter built in. Chapter 20 includes a handy circuit to build your own function generator.

Sweep generator A sweep generator is a special type of function generator; in most test gear, the sweep generator also includes a function generator. A sweep generator varies the frequency of the output waveform, typically within preselected limits, such as 100 Hz to 1 kHz, or 1 kHz to 20 kHz. Sweep generators are most often used in troubleshooting and aligning audio and video equipment.

Spectrum analyzer A spectrum analyzer looks a little like an oscilloscope, but it has a different function. The analyzer displays a small segment of the electromagnetic spectrum on a CRT tube. If there happens to be a radio signal in that segment, it is shown as a spike or "hill" in the readout. As radio signals most often consist of a main frequency plus numerous sidebands, the spectrum analyzer shows one tall spike, plus several others of diminishing heights on either side. Spectrum analyzers are most often used for designing and troubleshooting RF circuits, but they can also be used to analyze digital circuits. For example, as all high-speed digital circuits generate radio frequency emissions, it is sometimes helpful to take a closer look at the digital signals to see what effect the RF interference has.

Logic analyzer A logic analyzer is a souped-up oscilloscope. It shows the waveforms for several inputs and/or outputs of a single circuit. For instance, if you're testing an 8-pin logic chip, you can use a logic analyzer to show you the activity at each pin of the IC. The best logic analyzers have memory so you can "freeze" signals when they are at a particular state.

Signal injector A signal injector does as its name implies: it "injects" a signal, usually an audible tone, for the purpose of testing and aligning a circuit. A companion tool, a *signal tracer*, picks up the signal. You can readily build an audio frequency (AF) signal injector/tracer; plans appear in chapter 20.

Signal injector A signal injector does as its name implies: it "injects" a signal, usually an audible tone, for the purpose of testing and aligning a circuit. A companion tool, a *signal tracer*, picks up the signal.

4 Buying parts

TO ME, ONE OF THE JOYS OF EXPERIMENTING WITH ELECTRONICS IS BUYING parts for projects. Call me weird, but I enjoy the smell of a musty electronics shop, or the telltale odor of Bakelite in a well-stocked electronics surplus outlet. I spend hours perusing mail-order catalogs looking for the best deals in resistor and capacitor assortments.

From experience, I've found that the best way to simplify the construction of electronics projects is to use standard, off-the-shelf parts, things you can get at the neighborhood hardware store, auto parts store, and electronics store. This includes electronic components like resistors, capacitors, and transistors and also project boxes, motors, wire, knobs, and the other odds and ends that go into the typical project.

Finding parts for your projects is routine, all things considered, and little thought goes into it. Be forewarned, however, that there are some tricks of the trade, shortcuts, and tips that you should consider before you go on a buying spree and stock up for your next project. By shopping carefully and wisely, you can save both time and money in your electronics endeavors.

Places to Buy

There are numerous places to buy parts for your electronics projects. You'll find the best deals at:

- Hardware stores.
- Electronics stores.

- Speciality stores.
- Surplus stores.

Buying at hardware stores

Unless your electronic tinkering is limited to simple circuits built with few components, your projects will contain various hardware and construction items like angle brackets, sheet metal, wood, nuts, bolts, and so forth.

The small-town neighborhood hardware store is a great place when you need to buy some nails to fix the back porch or a rake to clean up the leaves scattered around your yard. But on closer inspection, it seems that while the hardware outlet carries thousands of items, it doesn't have the ones you need at that exact moment. The right size of angle bracket might not be stocked or is in short supply. The store might not carry a wide selection of extruded aluminum, and the exact one you want is nowhere to be found. Nuts and bolts are confined to prepackaged sets where they sell four bolts and five nuts—but you need five bolts and six nuts!

More often than not, you'll find it necessary to make trips to a variety of hardware stores, tapping the best that each has to offer. You might find that some stores cater to a specific segment of do-it-yourselfers and professionals. Some stores are designed expressly to please professional painters, while others are for weekend plumbers and electricians. Be aware of these specializations, and you'll have better luck in finding the parts you need.

Warehouse hardware outlets and builder's supply stores (usually open to the public) are the best source for the wide variety of tools and parts you need for advanced electronics work. Items like nuts and bolts are generally available in bulk, so you can save a considerable amount of money.

As you tour the hardware stores in your area, keep a notebook handy and jot down the lines each outlet carries. Then, when you find yourself needing a specific item, you have only to refer to your notes rather than bash your brain trying to remember where you saw some little doodad. On a regular basis, take an idle stroll during your regular hardware-store hunts. Unless the store is very small, chances are you'll find something new and laughably useful in electronic design each time you visit.

Buying at electronics stores

As recent as ten years ago, electronic parts stores used to be in plentiful supply. Even some automotive outlets carried a full range of tubes, specialty transistors, and other electronic gadgets. Now, Radio Shack remains as the only national electronics store chain. In many towns across the country, it's the only thing going.

Radio Shack continues to support electronics experimenters, but they stock only the most common components. If your needs extend beyond resistors, capacitors, and a few integrated circuits, you must turn to other sources. Check the local Yellow Pages under "Electronics—Retail" for a list of electronic parts shops near you.

Radio Shack isn't known for the best prices in electronic parts (although sometimes they do have really good bargains), but the neighborhood independent electronics specialty store might not be much better. It's not unusual for prepackaged resistors and capacitors to sell for 50 cents to $1 each, and few stores carry cost-saving parts assortments. Unless you need a specific component that isn't available anywhere else near you, stay away from the independent electronics outlet. There are independent stores that don't charge outrageously for parts, of course, but these are the exception, not the rule.

All is not lost. Mail-order provides a welcome relief to overpriced electronic components. You can find ads for these mail-order firms in various electronics magazines such as *Radio-Electronics* and *Modern Electronics*. Both publications are available at newsstands. A number of mail-order firms that cater to hobbyist electronics fans are listed in appendix A, "Sources."

Specialty stores

Specialty stores are those outlets open to the general public that sell items you won't find in a regular hardware or electronic parts stores. (Specialty stores don't include surplus outlets, which are discussed below.)

What specialty stores are of use to electronics hobbyists? Consider these:

- *Sewing machine repair shops* are ideal for small gears, cams, levers, and other precision parts. Some shops will sell broken machines to you. Tear the machine to shreds and use the parts for your projects.
- *Used-battery depot* is a good source for cheap batteries for your portable electronic gizmos. Most of the batteries are used, but reconditioned. The shop takes in old car and motorcycle batteries and refurbishes them. The selling price is usually between $15 and $25, or 50 to 75 percent less than a new battery.
- *Junkyards*. Old cars are good sources for powerful dc motors, used for windshield wipers, electric windows, and automatic adjustable seats. Bring tools to salvage the parts you want.
- *Bicycle sales/service shop*. Not the department store that sells bikes, but a *real* professional bicycle shop. Items of interest: control cables, chains, and reflectors.
- *Industrial parts outlet*. Some places sell gears, bearings, shafts, motors, and other industrial hardware on a one-piece-at-a-time basis. The penalty is high prices.

Shopping the surplus store

Surplus is a wonderful thing, but most people shy away from it. Why? If it's surplus, as the reasoning goes, it must be worthless junk. That's simply not true. Surplus is exactly what its name implies: extra stock. Because the stock is extra, it's generally priced accordingly—to move it out the door.

Surplus stores that specialize in new and used mechanical and electronic parts (not to be confused with surplus clothing, camping, and government equipment stores) are a pleasure to find. Most areas have at least one such surplus store; some as many as three or four. Get to know each and compare prices. Bear in mind that surplus stores don't have mass market appeal, so finding them is not always easy. Start by looking in the phone book Yellow Pages under "Electronics" and also under "Surplus."

Mail-order surplus

Some surplus is available through the mail. The number of mail-order surplus outfits that cater to the hobbyist is limited, but you can usually find everything you need if you look carefully enough and are patient. See appendix A, "Sources," for more information.

While surplus is a great way to stock up on electronics parts, odds and ends like knobs and panel meters, dc motors, and other goodies, you must shop wisely. Just because the company calls the stuff "surplus" doesn't mean that it's cheap. A popular item in a catalog could sell for top dollar.

Always compare prices of similar items offered by various surplus outlets before buying. Consider all the variables such as the added cost of insurance, postage and handling, and COD fees. Also, be sure that the mail-order firm has a lenient return policy. You should always be able to return the goods if they are not satisfactory to you. Some mail-order firms charge a restocking fee for returned items (the fee doesn't apply if the item is defective).

Make it a habit to purchase items you think you'll need in the near future, not just the ones that look "neat." Otherwise, all your money will go toward building an inventory you may never use—and believe me, I speak from experience here. If you have several projects in mind, jot down the parts you'll need for each, and order them at the same time. Turnaround time for shipping is usually under two weeks, even faster if you phone your order in and provide a credit card number for billing.

Bear in mind that most mail-order electronics firms impose a minimum order of $5 to $10, plus a shipping charge of $3 or more per order. That restricts you from ordering just a capacitor and couple of resistors at a time.

Avoid filling up the order to meet the minimum with items you don't really need. Instead, consider other projects you'd like to do, and get started ordering parts for them. If the part you need is available locally, it might cost less to buy it from a nearby electronics store than to pay shipping charges and worry about meeting the minimum order requirements.

Electronics Supplies and Components

Most of the electronic projects you'll assemble from this and other books with digital and analog circuits depend on a regular staple of common electronic components. If you do any amount of electronic circuit building, you'll want to stock up on the following standard components. Keeping spares handy prevents you from repeatedly making trips or placing orders to the electronics store.

Resistors

Get a good assortment of 1/4- and 1/8-watt resistors. Make sure the assortment includes a variety of common values (see Table 4-1), and that there are several of each value. Supplement the assortment with individual purchases of the following resistor values: 270 Ω, 330 Ω, 1 kΩ, 3.3 kΩ, 10 kΩ, and 100 kΩ.

The 270 Ω and 330 Ω values are often used with light-emitting diodes (LEDs) and the remaining values are common to TTL and CMOS digital circuits.

Table 4-1. Common Standard Resistor Values (ohms)

10	180	1.2 k	22 k
15	220	1.5 k	33 k
22	270	2.2 k	47 k
33	330	3.3 k	100 k
47	390	3.9 k	150 k
68	470	4.7 k	220 k
82	560	5.6 k	470 k
100	680	10.0 k	1 M
150	1 k	15.0 k	2.2 M

Variable resistors

Variable resistors, or potentiometers (pots) are relatively cheap and are a boon when designing and troubleshooting circuits. Buy an assortment of the small PC mount pots (about 80 cents each retail) in the 2.5 kΩ, 5 kΩ, 10 kΩ, 50 kΩ, 100 kΩ, 250 kΩ, and 1 MΩ values. You'll find 1-MΩ pots often used in op amp circuits, so buy a couple extra of these. I also like to have several extra 10 kΩ and 100 kΩ pots around since these find heavy use in most all types of circuits.

Capacitors

Like resistors, you'll find yourself returning to the same standard capacitor values project after project. For a well-stocked shop, get a dozen or so each of the following inexpensive ceramic disc capacitors: 0.1, 0.01, and 0.001 μF. The 0.1 and 0.01 μF caps are used extensively as *bypass* components and are absolutely essential when building circuits with TTL ICs. You can never have enough of these.

Many circuits use in-between values of 0.47, 0.047, and 0.022 μF. You might want to get a couple of these, too. Power supply, timing, and audio circuits often use larger polarized electrolytic or tantalum capacitors. Buy a few each of 1.0, 2.2, 4.7, 10 and 100 μF values. Some projects call for other values (in the picofarad range and the thousands of microfarads range). You can buy these as needed unless you find yourself returning to standard values repeatedly.

Table 4-2 lists the common standard values for capacitors.

Table 4-2. Common Standard Capacitor Values

pF			μF		
	2.0	100		0.001	0.47
	2.2	120		0.022	1.0
	3.3	150		0.033	2.2
	4.7	180		0.047	3.3
	10	220		0.068	4.7
	15	330		0.082	10
	22	470		0.010	22
	33	680		0.022	33
	47	820		0.047	47
	68			0.100	68
	82			0.220	100

Signal transistors

There are thousands of transistors available, and each one has slightly different characteristics than the others. However, most applications need nothing more than "generic" transistors for simple switching and amplifying. Common npn signal transistors are the 2N2222 and the 2N3904. Both kinds are available in bulk packages of 10 for about $1. Common pnp signal transistors are the 2N3906 and the 2N2907. Price is the same or a little higher.

By the way, you can often mix and match the plastic and metal can transistors, but it depends on the application. Noncritical circuits seldom care whether you are using a transistor in a plastic case or one in a metal case. The same is not true of some instrumentation and low-noise circuits, however, which often call for specific transistors, often in metal cases. For example, technically speaking, the plastic 2N2222 are actually called PN2222, while the metal-can version carries the 2N2222 designator. In any case, buy the plastic ones; they're cheaper.

If the circuit you're building specifies another transistor than the generic kind, you might still be able to use one of these if you first look up the specifications of the transistor called for in the schematic. A number of cross-reference guides provide the specifications and replacement equivalents for popular transistors.

Power transistors

There are common power transistors as well. The npn TIP31 and TIP41 are familiar to almost anyone who has dealt with power switching or amplification of up to 1 amp or so. Counterparts to the pnp are the TIP32 and TIP42. These transistors come in the TO-220 style of package.

A common, larger capacity npn transistor that can switch 10 amps or more is the 2N3055. It comes in the TO-3 style package and is available everywhere. The price is between 50 cents and $2, depending on the source.

Diodes

Common rectifier diodes are the 1N914, for light-duty signal switching applications, and the silicon 1N4000 series (1N4001, 1N4002, and so forth). Get several of each and use the proper size to handle the voltage in the circuit. Table 4-3 shows a quick rundown of common silicon rectifier diodes with their maximum average current and maximum peak reverse voltage ratings.

Table 4-3. Silicon Diode Specs

Number	Maximum Current (amps)	Maximum Peak Reverse Voltage (volts)
1N4001	1.0	50
1N4002	1.0	100
1N4003	1.0	200
1N4004	1.0	400
1N4005	1.0	500
1N4006	1.0	800
1N4007	1.0	1000
1N5400	3.0	50
1N5401	3.0	100
1N5402	3.0	200
1N5404	3.0	400
1N5406	3.0	600
1N5407	3.0	800
1N5408	3.0	1000
6A05	6.0	50
6A1	6.0	100

Table 4-3. Continued.

Number	Maximum Current (amps)	Maximum Peak Reverse Voltage (volts)
6A2	6.0	200
6A4	6.0	400
6A6	6.0	600
6A8	6.0	800
6A10	6.0	1000

LEDs

All semiconductors emit light (either visible or infrared), but light-emitting diodes (LEDs) are especially designed for the task. LEDs last longer than regular filament lamps and require less operating current. They are available in a variety of sizes, shapes, and colors. For most applications, the medium-sized red LED is perfect. Buy a few dozen and use them as needed. Some of the projects in this book call for infrared LEDs, which emit no visible light and are used in conjunction with an infrared-sensitive phototransistor or photodiode.

Although not all LEDs are created equally nor share the same specifications, the majority are general purpose and can be freely substituted without worry of voltage or current requirements. For the most part, you select an LED for a particular application based on its color, size, shape, and cost.

Integrated circuits

Integrated circuits let you construct fairly complex circuits from just a couple of components. Although there are literally thousands of different ICs, some with exotic applications, a small handful crops up again and again in hobby projects. You should keep the following ICs in ready stock:

- *555 timer.* This is by far the most popular integrated circuit for hobby electronics. With just a couple of resistors and capacitors, the NE555 can be made to act as a pulser, a timer, a time delay, a missing pulse detector, and dozens of other useful things. The chip is usually used as a pulse source for digital circuits. It is available in dual versions as the 556 and quadruple versions as the 558. A special CMOS version lets you increase the pulsing rate to 2 MHz.
- *LM741 op amp.* The LM741 comes second in popularity to the 555. The 741 can be used for signal amplification, differentiation, integration, sample and hold, and a host of other useful applications. The 741 is available in several dual versions such as the 1458 or the 747. The chip comes in different package configurations. The schematics in this book, and those usually found elsewhere, specify the pins for the common eight-pin DIP

package. Note that there are numerous op amps available, and some have design advantages over the 741.

- *TTL chips*. TTL ICs are common in computer circuits and other digital applications. There are many types of TTL packages, but you won't use more than 10 or 15 of them unless you're heavily into electronics experimentation. Specifically, the most common and most useful TTL ICs are the 7400, 7401, 7402, 7404, 7407, 7408, 7414, 7430, 7432, 7473, 7474, 74154, 74193, and 74244. You can find all or many of these in "TTL chip assortments," available through some of the mail-order electronics firms.
- *CMOS chips*. Because CMOS ICs require less power to operate than the TTL variety, you'll often find them specified for use with low-power applications. Like TTL, there is a relatively small number of common packages: 4001, 4011, 4013, 4016, 4017, 4027, 4040, 4041, 4049, 4060, 4066, 4069, 4071, and 4081.

Wire

Solid-conductor, insulated, 22-gauge hookup wire can be used both in your finished projects as well as for connecting wires in breadboards. Buy a few spools in different colors. Solid-conductor wire can be crimped sharply, so it might break when excessively twisted and flexed. If you expect that wiring in your project may be flexed repeatedly, use stranded wire instead. Heavier 12- to 18-gauge hookup wire is required for connection to heavy-duty batteries, motors, and circuit-board power supply lines.

Wire-wrap wire is available in spool or precut/prestripped packages. For ease of use, buy the more expensive precut stuff unless you have a tool that does it for you. Get several of each length. The wire-wrapping tool has its own stripper built in (which you must use instead of a regular wire stripper), so you can always shorten the precut wires as needed. Some special wire-wrapping tools require their own wrapping wire. Check the instructions that came with the tool for details.

Circuit boards

Simple projects can be built onto solder breadboards. These are modeled after the solderless breadboard, so you simply transfer the tested circuit from the solderless breadboard to the solder board. You can cut the board with a hacksaw or razor saw if you don't need all of it.

Larger projects require perforated boards. Get the kind with solder tabs or solder traces on them. You'll be able to secure the components onto the boards with solder. Most perf boards are designed for wire wrapping. Chapter 12 details the use of perf boards and solder breadboards.

Sockets

You should use sockets for ICs whenever possible. Sockets come in sizes ranging from 8-pin to 40-pin. The sockets with extra-long square leads (see Fig. 4-1) are for wire wrapping and when assembling a wire-wrapped project.

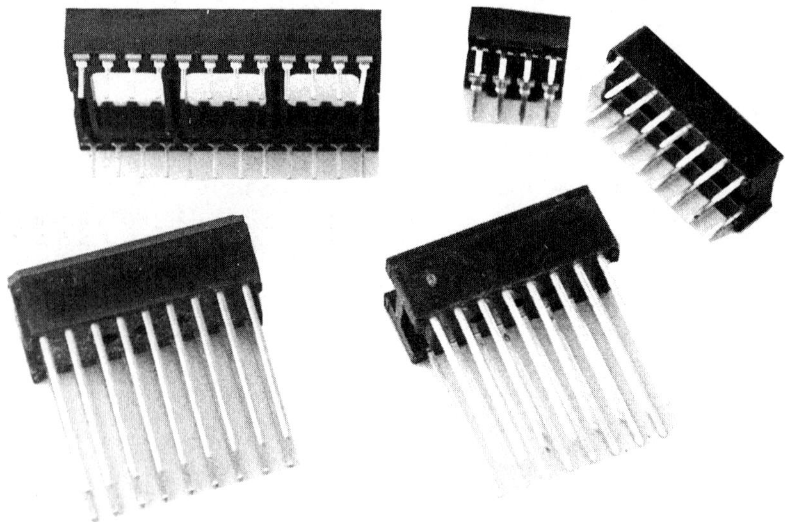

Fig. 4-1. An assortment of IC sockets.

You can also use wire-wrap IC sockets to hold discrete components like resistors, capacitors, diodes, LEDs, and transistors. You can, if you wish, wire wrap the leads of these components, but since the leads are not square, the small wire doesn't have anything to bite into, so the connection usually isn't very strong. After assembly and testing, when you are sure the circuit works, apply a dab of solder to the leads to hold the wires in place.

Cataloging Your Inventory

From time to time, you might want to catalog all the electronic components, hardware, and other paraphernalia in your stock. This helps to avoid shortages and prevents you from rebuying parts you already have.

Once the basic inventory is complete, it's a simple matter to maintain it by marking those components that you use. You can safely ignore common resistors, capacitors, LEDs, and diodes, but note when you use an IC or transistor. From time to time, review your inventory and look for items that have been depleted or are nearing depletion.

5

Electronics and safety

ELECTRONICS IS A SAFE AND SANE HOBBY, BUT ONLY WHEN PRACTICED WITH caution and respect. The construction and use of electronic devices can be dangerous to life and limb, and you must exercise reasonable care to make sure you live long enough (and with all of your eyes, fingers, and other body parts) intact to enjoy the projects you build.

Fortunately, the dangers of electronics are easily minimized by taking a few moments to recognize the jeopardy and steer away from the trouble spots. As detailed in this chapter, if you take your time and think things through, you never have to burn yourself with a soldering iron or shock yourself with an exposed electrical wire.

The Electronics Workshop Aptitude

To be a successful electronics hobbyist, you must be comfortable working with your hands and thinking problems through from start to finish. You should know how to use common shop tools and have basic familiarity in working with basic construction procedures. If you're just starting out in electronics, try a simple battery-powered project first, and work your way up. Don't attempt a complicated project that requires advanced construction skills, or high-voltage power.

With experience comes confidence, and with both come more professional results. Work on it long enough, and the electronic projects you build could be indistinguishable from expensive commercially made units.

You can learn a lot about electronics and the various skills that go into the science of electronics from this book (and the various other books that are written on the subject). Two important skills that you *can't* develop from reading books are *patience* and the *willingness to learn*. Both are absolutely essential if you want to build your own working electronic gadgets. Without these skills, you will most assuredly give up in desperation and frustration.

Give yourself time to experiment with your projects. Don't rush into things, because if so, you are bound to make mistakes. If a problem continues to nag at you, put the project aside, and let it sit for a few days or even a few weeks. You'll be surprised how the solutions to many of your most challenging problems will come when you least expect them (while dozing off for the night, showering, or while driving to work). Keep a small notebook handy and jot down your ideas so you won't forget them.

If trouble persists, perhaps you need to bone up on the subject before you can adequately tackle the problem. Take the time to study and learn more about the various sciences and disciplines involved. While you are looking for ways to combat the current dilemma, you are increasing your general electronics knowledge. Research is never in vain.

Working with ac

Many electronics projects operate from the typical 117-volt house current. A live ac wire can, and does, kill! Exercise caution whenever working with ac circuits. You can greatly minimize the hazards of working with ac circuits by following these basic guidelines:

- Always keep ac circuits covered.
- Keep ac circuits physically separate from low-voltage dc circuits. If necessary, construct a plastic guard within your project to keep the wiring separate.
- All ac power supplies should have fuse protection on the incoming hot line. The fuse should be adequately rated for the circuit but should allow a fail-safe margin in case of a short circuit.
- When troubleshooting ac circuitry, keep one hand in your pocket at all times. Use the other hand to manipulate the voltmeter, oscilloscope probe, or other testing apparatus. Avoid the situation where one hand touches ground and the other a live circuit. The ac current will flow from one hand to the other, through your heart.
- When possible, place ac circuits in an insulated Bakelite or plastic chassis. Avoid using a metal chassis.
- Double- and triple-check your work before applying power. If you can, have someone else inspect your handiwork before you switch the circuit on for the first time.

- Periodically inspect ac circuits for worn, broken, or loose wires and components, and make any repairs necessary.

While you can build dc-operated electronics projects that lack a suitable enclosure, *all* ac circuits must be adequately covered during operation. The exception is when actually testing the circuit. Most accidents occur during this testing phase, so be particularly alert.

If possible, test the circuit with power removed. This can often be accomplished with the use of a logic probe, a signal injector, and other passive testing equipment. When testing ac-operated circuits with power removed, *unplug* the power cord—don't just switch off the power at an outlet strip. You can always be certain the power is disconnected when the plug has been pulled from the socket.

High-Voltage Electrocution

A number of advanced electronic circuits use voltages higher than 117 Vac found at wall outlets. For example, a Tesla coil apparatus generates 20,000 volts or more. A power supply for a helium-neon laser produces 1,200 to 3,000 volts. Other projects, like Geiger counters, stun guns, Van de Graaff generators, night vision scopes, and bug zappers produce in excess of 1,000 volts. These voltage potentials should never be taken lightly, especially when you consider that the electric chair puts out only 2,000 volts!

Some electronics enthusiasts tend to disregard the high voltages often found in projects such as laser power supplies and Tesla coils. They assume that although the voltage is high, the current level is low. They rely on the old maxim "It's the volts that jolts, but it's the mills that kills." This is true but limited thinking. For example, the current demand of the typical helium-neon laser is low, between 3 and 7 milliwatts. At normal dc levels (5 or 12 volts), this current is not enough to be felt. But at the 1.2 to 3 kV level of standard laser operating voltages, even a seven milliwatt jolt can, depending on the circumstances, kill you.

High-voltage power supplies should be properly shielded and insulated. Avoid operating a power supply in the open and always cover exposed high-voltage parts. At about 300 volts, electricity begins to ionize the surrounding air, and arcing can occur. You can receive a bad shock from a high-voltage component or power supply without even touching anything. A 3 kV arc can jump a quarter of an inch or more, and like lightning, discharge on the nearest object. If that object happens to be your fingers or elbow, you will receive a painful shock.

Insulating material such as high-voltage putty (available at TV repair shops and electronic stores) restricts arcing and provides a relatively shock-proof layer. It's easy to forget that several thousand volts are coursing through a wire, and you might inadvertently touch it or brush across it.

Admittedly, most shocks from low-current, high-voltage power supplies (like those used with helium-neon lasers, neon sign transformers, and so forth) won't kill you, nor will they burn your skin. But don't underestimate high voltages, no matter what the current level. You body's protective mechanism automatically reacts to the jolt. If you touch a live wire with your hand, your body quickly contracts to prevent further shock. If the jolt is large enough, you may be knocked backward onto the ground.

Phantom current

Many high-voltage power supplies use a series of capacitors and diodes to multiply the incoming voltage to higher levels. The capacitors tend to retain some current even after you've switched the power supply off. This is "phantom current" and is well known to anyone who has worked with TVs, where the TV tube can maintain a high voltage charge for weeks or even months since the set was last turned on.

When working on high-voltage power supplies, be sure to short the output wires together to deplete any remaining charge in the capacitors. Do this even if the power supply uses "bleeder" resistors to drain the capacitors. Use the circuit shown in Fig. 5-1 to short the positive and negative output terminals together. The resistors prevent the capacitors from discharging too quickly. This helps prolong the life of the capacitors. Keep the shorting wire in place for a second or two until all the current has been drained.

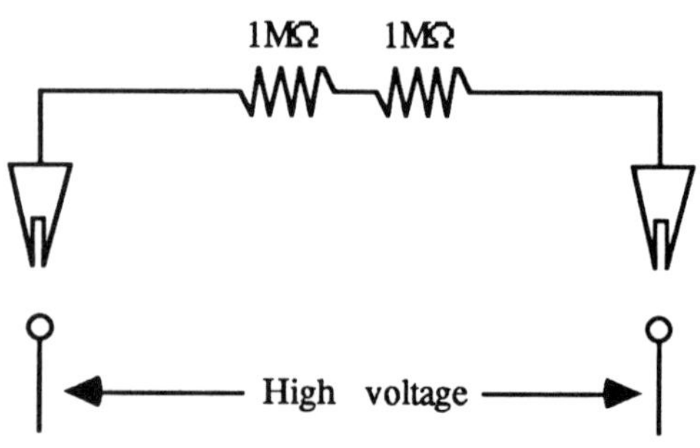

Fig. 5-1. Shorting wire for high-voltage output.

High-voltage precautions

Take these simple steps to avoid unnecessary high-voltage shocks:

- Do not apply power to the project until all the high-voltage wires and components are shielded or insulated.
- Do not work near a project that has high-voltage components exposed.
- Provide an interlock switch to protect against accidental electrocution when servicing a high-voltage device.

- Affix a "Danger—High Voltage" sticker, like the one in Fig. 5-2, to the power supply to warn others that high voltages are present.
- Install a power key switch to prevent unauthorized use.
- Never leave a high-voltage device or power supply unattended. For maximum security, lock the project in a desk or filing cabinet to prevent unauthorized use.

Fig. 5-2. A sticker for placing in areas that could pose a hazard to user or technician.

Safe Soldering

There are many electronic construction techniques including breadboarding, wire wrapping, and soldering. Soldering requires that you use a hot soldering iron or "pencil," working with temperatures exceeding 600 degrees F. This temperature is equivalent to an electric stove burner set at medium-high heat, so you can imagine the dangers involved.

Electronics requires the use of soldering pencils or small irons. The older soldering guns create too much heat in too large of an area. Your soldering iron should be outfitted with the proper tip for the job. Avoid a tip that's too small or too large for the work. A small tip won't produce enough localized heat, requiring that you keep the iron in one place for an excessive period of time. Conversely, a large tip produces too much heat, causing potential damage to components.

The printed circuit board tip, with its small bevel, is used for most applications where you need to solder components to each other or to a printed circuit board (this includes the "universal solderboards" detailed in chapter 12, "Using Experimenter's Breadboards and Solderboards").

Always place your soldering iron in a stand designed for the job. *Never* place a hot iron on the table. You could easily start a fire or burn your hands. Be certain that the iron's electrical cord does not snag on the table or any other object. Otherwise, the hot iron could get yanked out of its stand and fall onto your work, or worse, right into your lap.

Mildly caustic and toxic fumes are produced during soldering. Maintain good ventilation to prevent a buildup of these fumes in your working area. Avoid hunching over the soldering work too much, as the fumes will waft into your face. If you're having trouble seeing the soldering joint, use a magnifying glass to bring the *image* of the work closer to you.

Fire Control

Fire is a potential hazard during the construction and use of electrical devices. A hot soldering iron can easily ignite paper, wood, and cloth. A short circuit can cause a spark, which could ignite a flammable object, including your clothes. Although less common, an electronic circuit can develop too much heat and melt or burn its enclosure and surroundings. Proper construction techniques and careful review of your work should prevent many mishaps. Always build and operate your electronics projects away from flammable objects, including gasoline, lighter fluid, welding and brazing equipment, and cleaners.

If your project operates under 117 Vac house current, keep an eye on it for the first several hours of operation. Note any unusual behavior, including arcing, overheating, or circuit burnouts. If a circuit breaker trips while your project is on, you can bet that something is amiss with your wiring.

Always keep a fire extinguisher near you, and don't hesitate to use it if a fire should break out. The Halon-type fire extinguisher uses a colorless gas to snuff out the flames. These are preferred over the standard powder-type extinguishers because they don't make as much of a mess. Bear in mind that Halon works by displacing oxygen. *Immediately vacate* the room after using the fire extinguisher, or you could succumb to oxygen deprivation. After the fire is out, air out the room completely for at least 15 minutes before reentering.

On an environmental level, Halon is a potent chlorinated hydrofluorocarbon and therefore disrupts the high-altitude ozone layer that protects the earth against harmful ultraviolet radiation from the sun. Odds are, however, that you'll never have to use the fire extinguisher, and the relatively small amount released during your fire-fighting effort will not overly contribute to the ozone layer problem.

Projects Safety

Some of the projects you build might require you to work with potentially hazardous materials including chemical etchants, glass, and wood and metal tools. Glass and mirror cutting should be done only with the proper tools and safety precautions. Wear eye protection whenever you are working with glass. If you can, wear gloves when handling cut glass and mirror. Finish the edges with a burnisher, flame, or piece of masking tape.

Superconductors and certain other electronics experiments involve the use of liquid nitrogen, a cryogenic gas that has a temperature of minus 196 degrees Centigrade. While liquid nitrogen is nonflammable and nontoxic, its extremely cold temperature can cause frostbite burns. Handle liquid nitrogen only in an approved container (a Thermos bottle with a hole drilled in the cap often works), waterproof gloves, and eye protection.

Many projects require the use of wood and metal working tools for fashioning project cases, bases, and other components. You can use hand or power tools; either way, carefully follow all operating procedures and use the tools with caution. Most power tools provide some type of safety mechanisms. Never defeat them! Thoroughly read the instruction manual that came with the tool.

First Aid

Despite your best efforts, accidents can and do happen. Most will be minor, causing little or no injury. If injury does occur, be sure to treat it promptly. If necessary, see a doctor to prevent the condition from getting worse.

Consult any book on medical first aid treatment for details on how to care for cuts, abrasions, and other minor injuries. You'll want to keep a first aid kit handy at all times, preferably right there in your shop, or conveniently located in a wash room or lavatory. Periodically check the contents of the first aid kid and look for:

- An ample supply of self-adhesive bandages for minor cuts.
- Sterile gauze and adhesive tape for larger cuts and lacerations.
- Tweezers, for picking out splinters of wood and glass, or clipped metal component leads.
- Scissors, for cutting gauze and bandages.
- Isopropyl alcohol.
- Sterilized disinfectant, such as Betadine (this is regularly available in squeeze bottles at drug stores).
- Hydrogen peroxide.
- Miscellaneous supplies including cotton-tipped swabs, sterilized ointment, aspirin, etc.

Some cleaners and disinfectants, including hydrogen peroxide, have a limited shelf life. These should be inspected and replaced on a regular basis. Prior to applying any disinfectant or bandage, be sure the injury is completely cleaned, using a mild soap (preferably a sterilized soap). This is particularly important with deep cuts or puncture wounds.

Minor burns from soldering irons or hot components are usually best treated by leaving them alone—many doctors and nurses do not recommend application of creams, salves, or bandages to minor burns. More serious burns might require proper medical treatment.

Chemical burns, such as those from circuit-board etchants or battery acid, should be treated by a doctor. The chemical must be completely removed or neutralized before treating the actual burn.

Perhaps most serious of all is injury to eyes. You might want to consider wearing unbreakable goggles when assembling your projects. Chips of solder and metal leads can fly off when they are cut, and there is a chance that harmful debris could strike your eye. Should you get something in your eye, especially a piece of glass, metal, or plastic, see a physician *immediately*. Trying to remove the object yourself can cause further damage.

If you are like most electronics hobbyists, you'll receive at least one shock from a live 120 Vac circuit. Odds are, it'll shake you up and make you jump around the room a little bit (old-time electricians call this the "kilowatt dance"). Your heart might speed up a bit or even skip a beat, but otherwise, you are likely to pull through with no ill side effects.

Usually.

Check the area of skin contact for signs of burns. Treat it as you would any other burn. Immediately after an electrical shock, stop what you are doing and consciously calm yourself down. Monitor your pulse to make sure your heart isn't suffering from arrhythmia (uneven beating). If you feel anything is amiss, consult a doctor immediately. Seek medical attention if you feel dizzy or light-headed or if you feel a pain in your chest following a bad shock.

It's always a good idea to experiment with electronics using the "buddy system." Enlist the help of a family member or friend to keep an eye out for you while you're poking around inside electronic circuits. Should you receive a bad shock and become unconscious, this buddy could mean the difference between your life and death. Take the time to learn and practice CPR and keep emergency numbers handy at all times.

Let Common Sense Prevail

Common sense is the best shield against accidents, but common sense can't be taught or written about in a book. It's up to you to develop common sense and use it at all times. Never let down your guard. Don't ruin the fun of a wonderful hobby or vocation because you neglected a few safety measures.

6

Eliminating static electricity

THE ANCIENT EGYPTIANS DISCOVERED STATIC ELECTRICITY WHEN THEY rubbed animal fur against the smooth surface of amber. Once the materials were rubbed together, they tended to cling to one another. Similarly, two pieces of fur that were rubbed against the amber tended to separate when they were drawn together. While the Egyptians didn't understand this mysterious unseen force, they knew it existed.

Ben Franklin and many others of his time understood more about static electricity. One of Franklin's many inventions is an early motor that runs completely on static electricity. Franklin's motor consists of a spoked wheel with a metal thimble on the end of each spoke. Two Leyden jars, which hold positively and negatively charged static electricity, provide the power. The motor turns as the static flows from the jars and onto the thimbles. The action of electrical attraction and repulsion keeps the motor spinning until the static in the jars is depleted.

Today, you can encounter static electricity by doing nothing more than walking across a carpeted floor. As you walk, your feet rub against the carpet, and your body takes on a static charge. Touch a metal object, like a doorknob or a metal sink, and that static is quickly discharged from your body. You feel the discharge as a shock.

Carpet shock has never been known to kill anyone. The amount of voltage and current is far too low to cause great bodily harm. But the same isn't true of electronic circuits. Considering how your body might develop a 10,000- to

50,000-volt charge when you walk across a carpet, you can imagine what that might do to electrical components that are rated at just 5 or 15 volts. The sudden crash of static can burn holes right through a sensitive transistor or integrated circuit, rendering it completely useless.

As an electronics experimenter, you will want to take specific precautions against electrostatic discharge (ESD). Damage from static discharge can be practically eliminated by taking just a few simple steps to protect you, your tools, and your projects from static buildup. The cost for protecting against static electricity is minimal; without knowing it you may already be on the road to preventing dangerous static buildup in your electronics workshop.

The Problem of Electrostatic Discharge

Electrostatic discharge involves very high voltages at extremely low currents. Combing your hair on a dry day can develop tens of thousands of volts of static electricity, but the current is almost negligible. The low current prevents you from injury when you receive a shock from a static discharge.

Many semiconductor devices are not so forgiving. Transistors and integrated circuits designed around a metal-oxide substrate can be particularly sensitive to high voltages, regardless of the current level. These components include MOSFET transistors, CMOS integrated circuits, and most computer microprocessors. Other electronic components are also sensitive to very high levels of electrostatic discharge, but these levels are not normally encountered in everyday life.

A good number of projects you build might include MOS transistors and CMOS chips, so you'll need to consider the problems of electrostatic discharge. Note that today's MOS and CMOS devices are far less sensitive to the destructiveness of static than they used to be. However, a certain level of preventative care is still required. The preventative measures involve grounding the components, yourself, and your tools.

Storing MOS and CMOS Devices

Plastic is one of the greatest sources of static electricity. Rub a piece of plastic, especially thin plastic like food wrap, against just about any material, and you'll develop a healthy amount of static electricity.

Storage and shipping containers are often made of plastic, and it's a great temptation to dump your MOS and CMOS devices into these containers. Don't do it. Invariably, static electricity will be developed and the component could be damaged.

Unfortunately, there's no way to tell if a MOS or CMOS part has become damaged by electrostatic discharge just by looking at it, so you won't know things are amiss until you actually try to use the component. At first, you might think the circuit has gone haywire or that your wiring is at fault. If you're like most, you won't blame the transistors and ICs until well after you've torn the rest of the circuit apart.

CMOS chips and MOS transistors are best stored using one of the following methods. All work by grounding the leads of the IC or transistor together, thereby diminishing the effect of a strong jolt of static electricity. Note that none of the storage methods is 100 percent foolproof.

Antistatic mat This mat looks like a black sponge, but it's really conductive foam. You can (and should) test this by placing the leads of a volt-ohmmeter on either side of a length of foam. Dial the meter to ohms. You should get a reading instead of an open circuit. The foam can be easily reused, and large sheets make convenient storage pads for many components. Some antistatic mats are pink in color. Beware of assuming that a pad is antistatic just because it's black or pink. Test it first using a volt-ohmmeter. If you are storing your components in a project cabinet, as shown in Fig. 6-1, you can cut the antistatic foam to size to exactly fit inside the drawers.

Fig. 6-1. Cut antistatic foam to size and use it to line the insides of your parts cabinet drawers.

Antistatic pouch or bag Antistatic pouches are made of a special plastic that generates little static and are coated on the inside with a conductive layer. The bags are available in a variety of forms. Many are a smoky black or gray color, others are pink or jet black. As with mats, you should never assume a storage pouch is antistatic just from its color. Check the coating on the inside with a volt-ohmmeter.

Antistatic tube Large quantities of chips are most often shipped and stored in convenient plastic tubes. These tubes help protect the leads of the IC and are well suited to automatic manufacturing techniques. Ordinary ICs use standard plastic tubes; CMOS integrated circuits use antistatic tubes. The construction of the tube is similar to the antistatic pouch: plastic on the outside and a thin layer of conductive material on the inside.

Remove the chip or transistor from its antistatic storage protection only when you are installing it in your project. The longer the component is protected, the better.

MOS transistors can often be stored without an antistatic mat or pouch by wrapping a thin, uninsulated wire around all of its leads. A length of stripped, 30 AWG wire-wrapping wire is a good choice. Wrap the ends around the tips of the leads so you can remove the wire when it comes time to install the transistor. For best results, remove the wire completely only when the transistor has been soldered into place or inserted in its socket on the circuit board.

Grounding Your Workbench

An electrostatic charge can be developed when building projects as well. Handling plastic project boxes, using tools, or scuffing your feet on the carpeted floor can generate thousands of volts of static electricity. If you plan on experimenting with electronics to any great extent, you'll want to consider providing an earth ground to the top of your workbench.

You can ground your benchtop by attaching an insulated wire to a piece of thin, bare, sheet metal measuring about 12 by 18 inches. You can use almost any metal, including stainless steel or aluminum, but it must be bare—not painted or varnished—and it should resist rust and corrosion. Attach the other end of the wire to a 1 MΩ, 1/2-watt resistor, as shown in Fig. 6-2. The other end of this resistor is securely attached to an earth ground, such as a cold (not hot) water pipe.

You can also tap into the grounding lug in a nearby electrical outlet, but be sure you attach the wire to the ground lug (see Fig. 6-3) and not to the hot or neutral side of the receptacle. Otherwise, you could receive a nasty shock.

Do not omit the resistor. It is there as a safety precaution and could save your life. Omitting the resistor could bring you in direct contact with earth ground. Should you touch a live ac wire at the same time, you (actually your

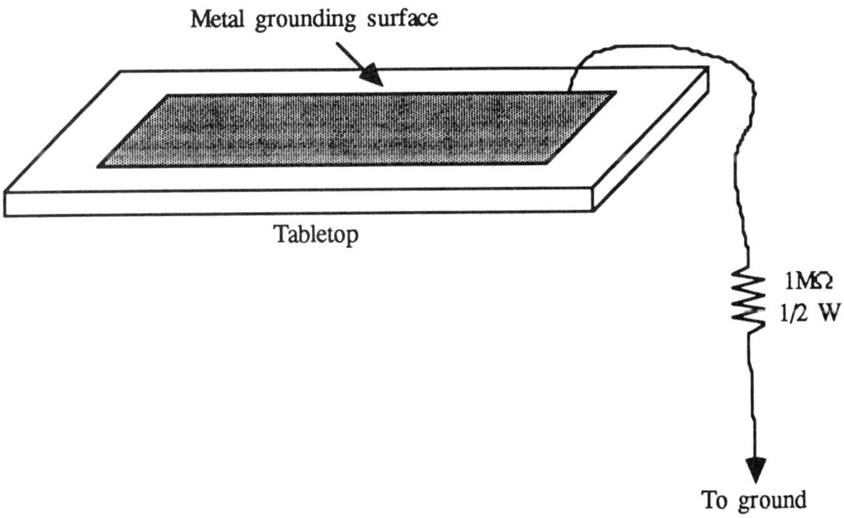

Fig. 6-2. How to safely ground the top of your workbench. *Do not* leave out the resistor.

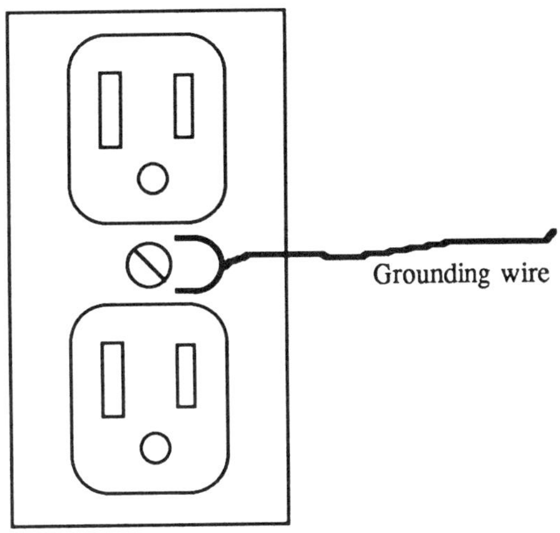

Fig. 6-3. How to attach a grounding wire to an ac receptacle. Test the receptacle first to make sure it is properly wired.

heart) will be in the middle of 120 volts (or possibly more). Touching a live wire and ground at the same time is often fatal.

As another approach, you can purchase resistive work mats from most electronics supply stores. The mats come in various sizes and not only provide a cushion that protects your tools and projects against scrapes and mars, but they provide the necessary ground to help eliminate harmful electrostatic discharge. Price can be steep, so shop around. Buy a pad large enough for the job. Pads are also available for under your table to prevent you from building up a static charge from scuffing your feet. Remember that no matter what type or size of pad you use, it is useless if it is not properly grounded.

Using an Antistatic Wrist Strap

As a further aid in reducing static electricity, wear an antistatic wrist strap when working with MOS and CMOS devices. This wrist strap, like the one shown in Fig. 6-4, grounds you at all times and prevents static buildup. The strap is one of the most effective means of eliminating electrostatic discharge, and it's one of the least expensive. The wrist strap pictured in the illustration was purchased at Radio Shack for under $3.

To use the strap, roll up any shirt sleeves and remove watches, bracelets, or other metallic objects. Wrap the strap around your wrist and make sure it's tight. Most straps use Velcro or similar material to make fastening easier. Then securely attach the wire from the wrist strap to a proper earth ground.

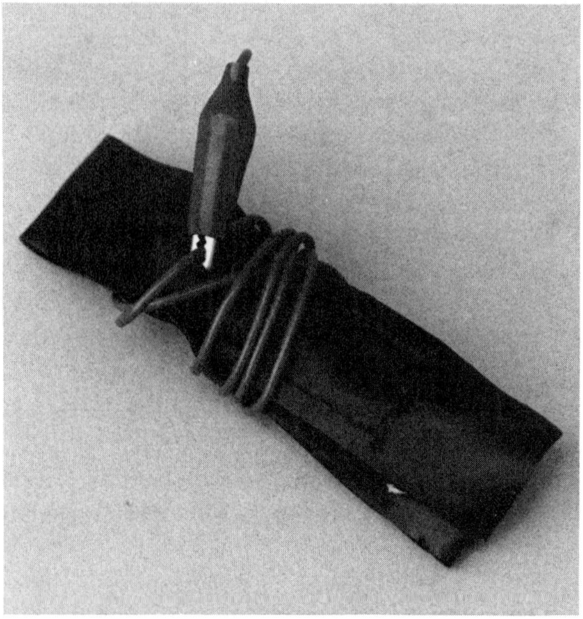

Fig. 6-4. An antistatic wrist strap.

Proper Clothing for Static Reduction

Your choice of clothing can affect the amount of static buildup in your body. Whenever possible, wear natural fabrics, such as cotton or wool. Avoid wearing polyester and acetate clothing, as they have a tendency to develop copious amounts of static.

You can wear a cotton lab overcoat to help reduce static electricity. Overcoats are available at many chemical and industrial supply houses. You can also find suitable overcoats, smocks, and aprons at many hardware and wood supply stores.

Grounding Tools

If your soldering iron operates from ac current, it should be grounded, whether or not you are using MOS and CMOS devices. A grounded iron not only helps prevent damage from electrostatic discharge, but it lessens the chance of a bad shock should you accidentally touch a live wire.

Unfortunately, the less expensive soldering irons are two-prong only, meaning they lack a ground connection. There really is no safe means of attaching a grounding wire to the iron, so the best bet is to just buy a new one. You can purchase a grounded soldering iron for less than $30, including an assortment of tips. The ungrounded models cost a few dollars less, but they have no place in the electronics workshop when using modern MOS and CMOS components.

As long as you are grounded by using a antistatic wrist strap, there is generally no need to ground your other tools such as wire-wrapping tool, screwdriver, and wire cutters. However, these tools can be grounded by attaching wires to them and leading the wires to a suitable ground. Again, use the 1 MΩ, 1/2-watt resistor to prevent a direct path from the tool to ground. You might want to devise a clip-on grounding clamp that attaches to your tools. Before using the tool, attach the clamp securely to the exposed metal of the tool (scrape away paint and insulation, if necessary).

Using Sockets

Commercially manufactured electronic products seldom use sockets for their integrated circuits, but your own homemade projects should use sockets for no reason other than to avoid damage from electrostatic discharge. When building your projects that use ICs, install sockets first. When the entire circuit has been completely wired, you can check your work, and then add the chips. This technique should be used whether the integrated circuits are CMOS or TTL because the sockets provide a handy means of replacing or reinserting the chips.

Sockets are available in all the various IC sizes, from 8-pin to 40-pin and beyond. Stock up on a selection of 8-, 14-, and 16-pin sockets. The larger 20-, 24-, and 40-pin ICs and sockets are found primarily in computer circuits, although there are plenty of large ICs used in other applications.

Sockets can also be used for transistors, though most bipolar and power transistors are not overly sensitive to electrostatic discharge.

7

Schematics and electronic symbols

ELECTRONIC DEVICES USE SPECIALIZED ROAD MAPS TO TELL YOU WHAT COMPONENTS are used and how they are connected together. This pictorial road map is the *schematic*, a kind of blueprint that tells you just about everything you need to know to build an electronic circuit.

Schematics are composed of special symbols with interconnecting lines. The symbols represent individual components, and the lines are the wires that connect these components together. The language of schematics, while far from universal, is designed to enable duplicating the construction of a circuit with little more information than that picture.

The experienced electronics experimenter knows how to read a schematic. This entails recognizing and understanding the various symbols used to represent electronic components and how these components are connected. All in all, learning to read a schematic is not difficult; with previous knowledge, you can learn to read a basic schematic in less than 10 minutes.

Part of the difficulty in reading schematics is remembering each of the symbols used to represent components. There are literally hundreds of symbols commonly used in today's electronic circuits, and even more exist in older circuits that use tubes and other non-solid-state devices.

In reality, however, you need only be familiar with a dozen or so symbols; the rest you can look up. This chapter details several dozen electronic symbols including those for elementary components like resistors and capacitors, logic symbols like the AND and OR gate, transistors, and more. The symbols are listed alphabetically within the following categories:

- Basic schematic symbols: chassis and earth ground, connection points, inputs, and outputs.
- Electronic components: resistors, transistors, diodes, and chokes.
- Logic symbols: AND, OR, NOR, and inverter gates.
- Miscellaneous symbols: switches, light bulbs, and other "hardware" items.

Basic Schematic Symbols

Connections

Phono/coaxial

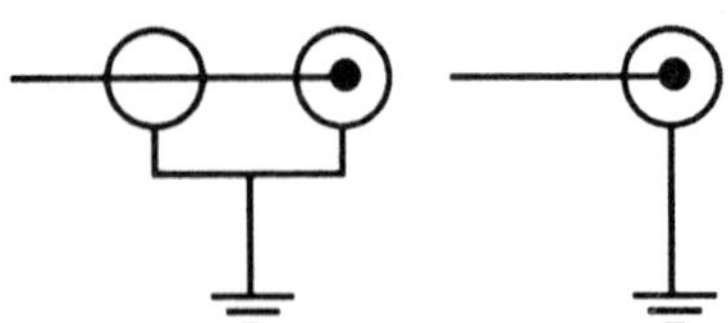

Jack Plug

Concentric/phone
2 conductor

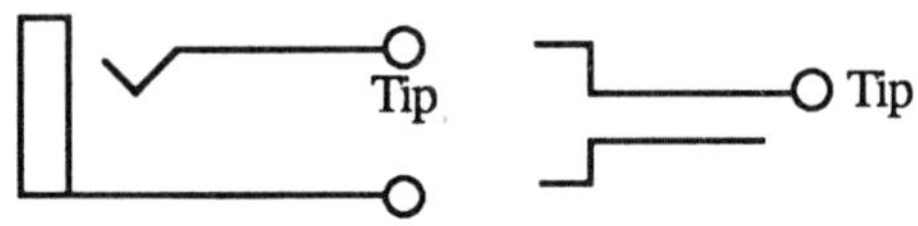

Jack Plug

Concentric/phone
3 conductor

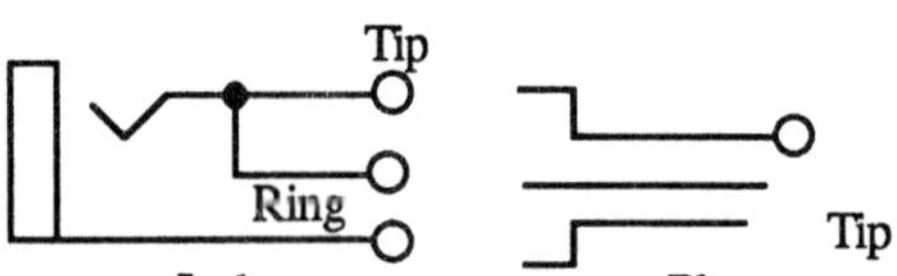

Jack Plug

Concentric/phone
3 conductor
Switch contact jack

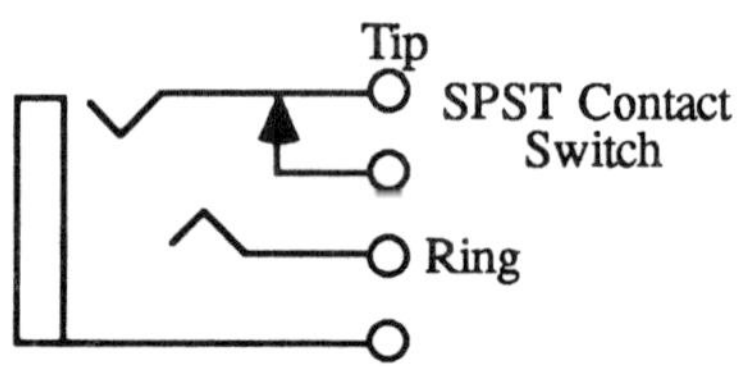

Chassis disconnects

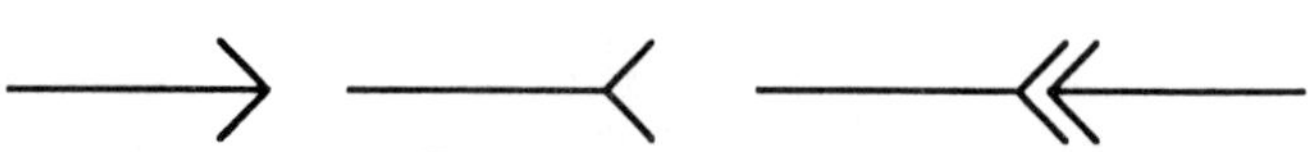

Male Female Engaged

Terminals

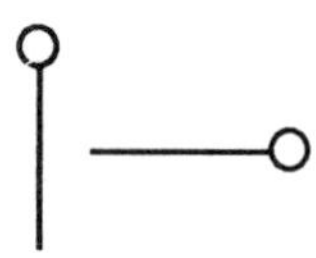

Test points

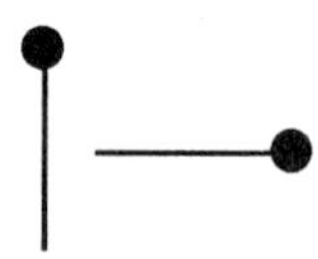

Grounds

Earth ground

Chassis ground

Note: Earth ground schematic now commonly used in books and magazines for denoting chassis ground

Single common

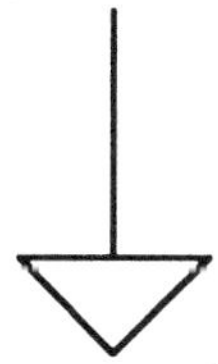

Multiple common (# denotes number)

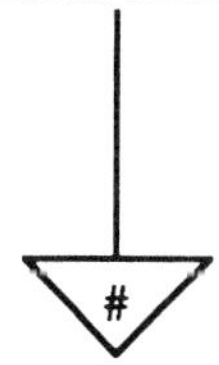

Common tie points

Wires

Wire connections

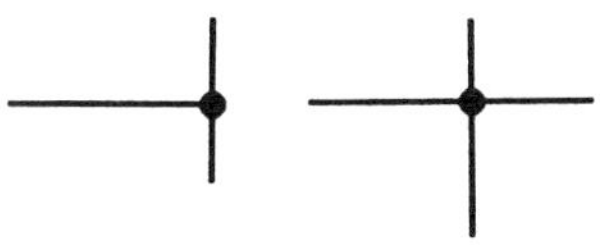

Connected

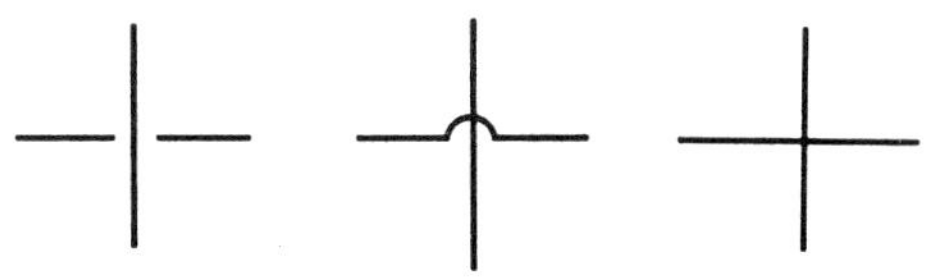

Not connected

Shielded

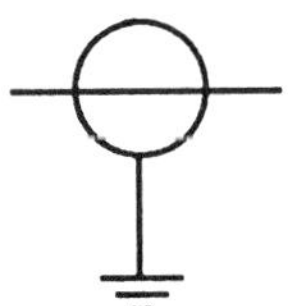

Shielded w/ Ground

Shielded

Multiple data lines

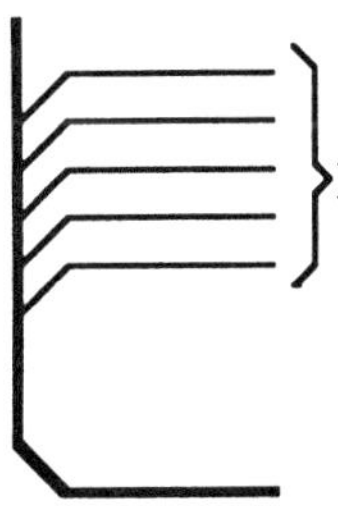

Data lines connecting at bus

Data bus

Electronic Components

Bridge rectifier (see Diodes, bridge)

Capacitors

Fixed; nonpolarized

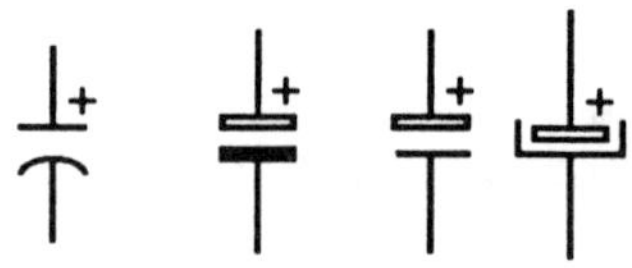

Fixed; polarized

Variable

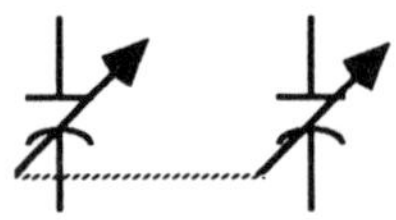

Variable; ganged

Voltage variable (varactor)

Choke (see Inductors)

Crystal

Diac

Diodes

Rectifier

Zener

Pin

Tunnel

Bridge

Light-Emitting Diode (LED)

Photodiode

Inductors

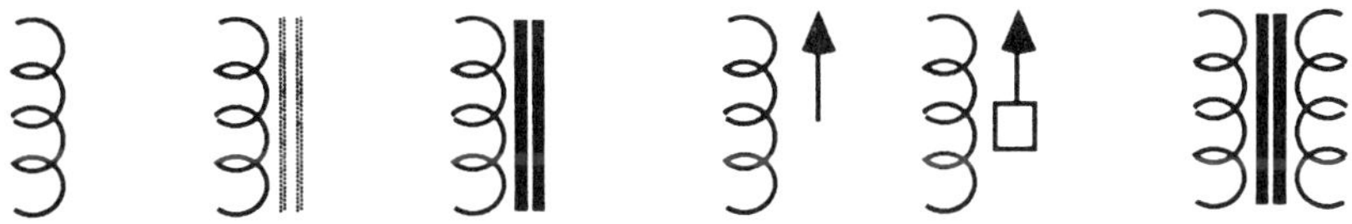

Air core Powdered-iron core Iron core Variable core Transformer

Operational amplifier

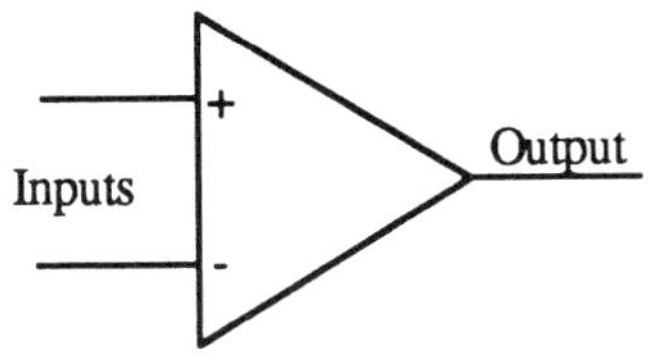

Potentiometer (see Resistors, variable)

Rectifier (see Diodes, rectifier)

Relay

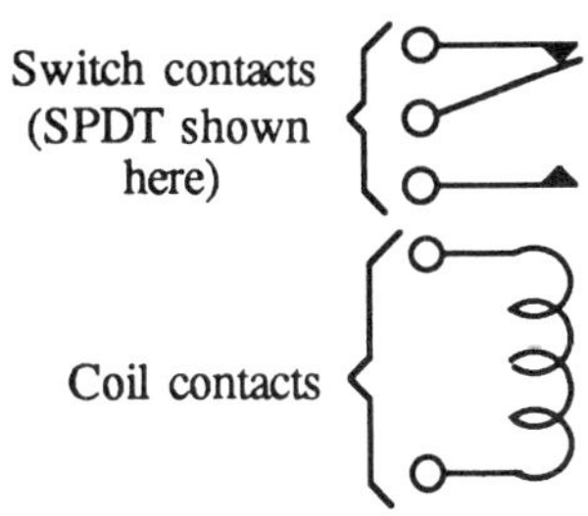

Resistors

Fixed Variable (Potentiometer) Variable (Potentiometer) Voltage-dependent Current-dependent

Light-dependent Temperature-dependent

Transistors

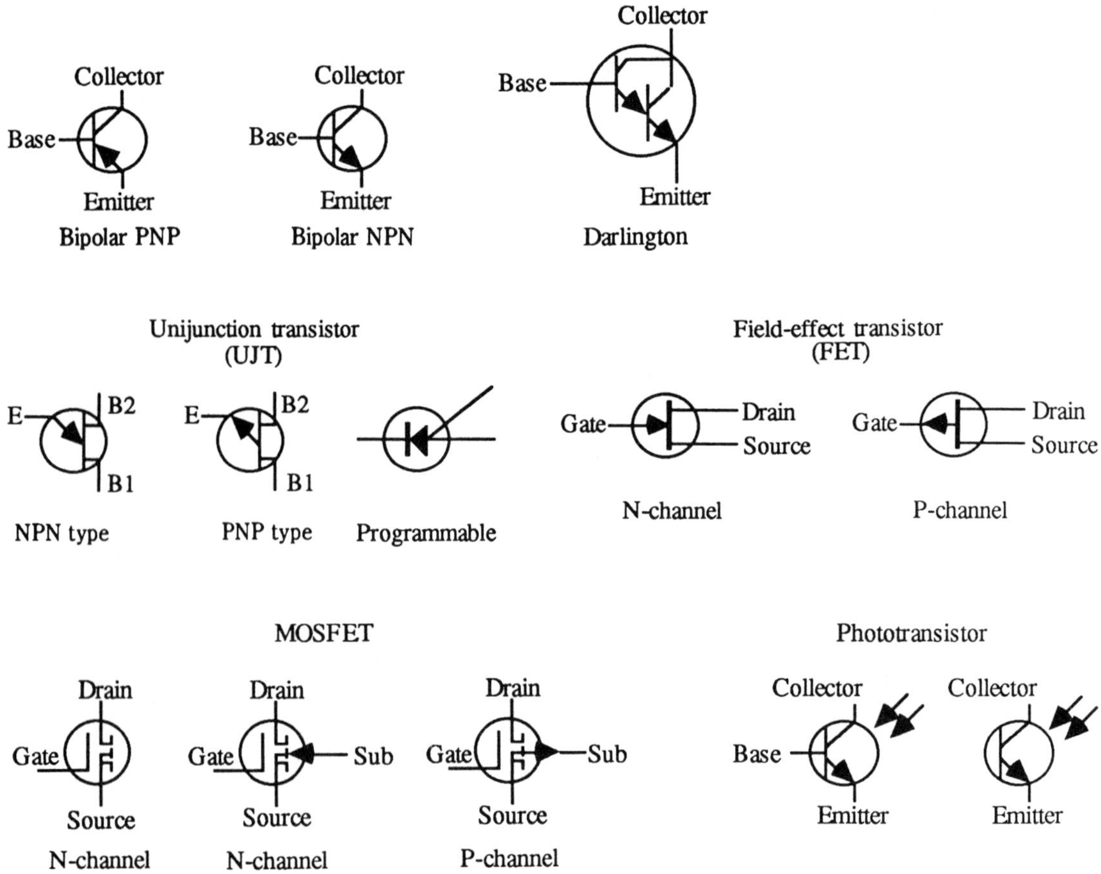

Transformers

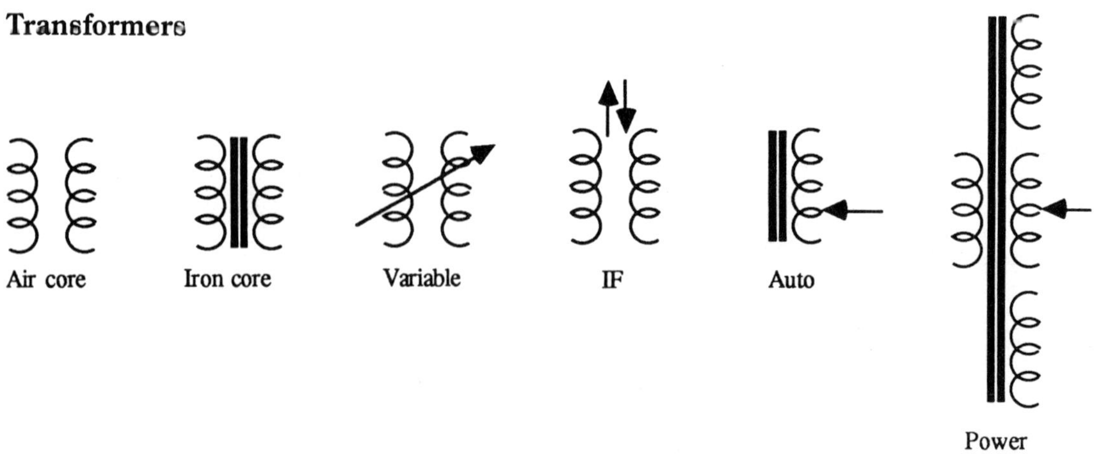

Triac

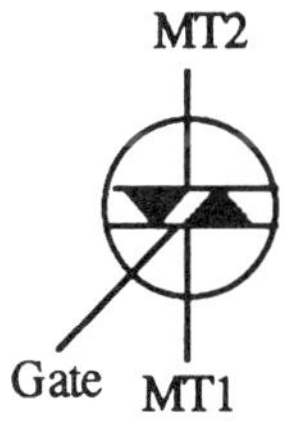

Zener diode (see Diodes, zener)

Logic Symbols

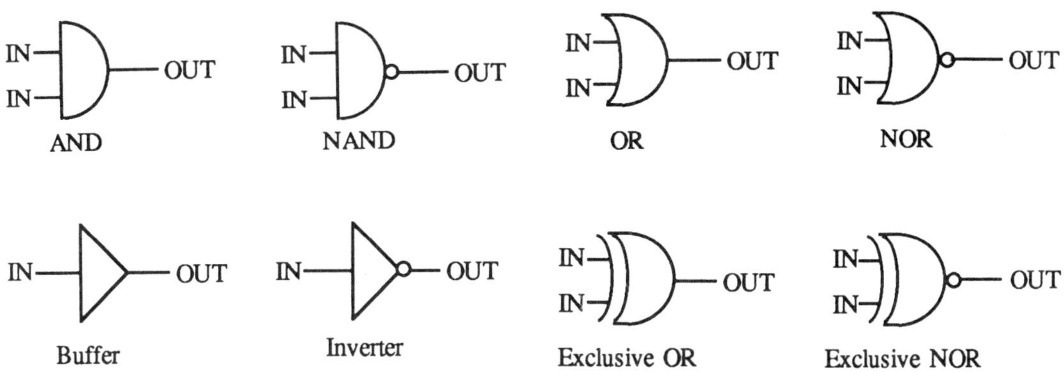

Flip Flops

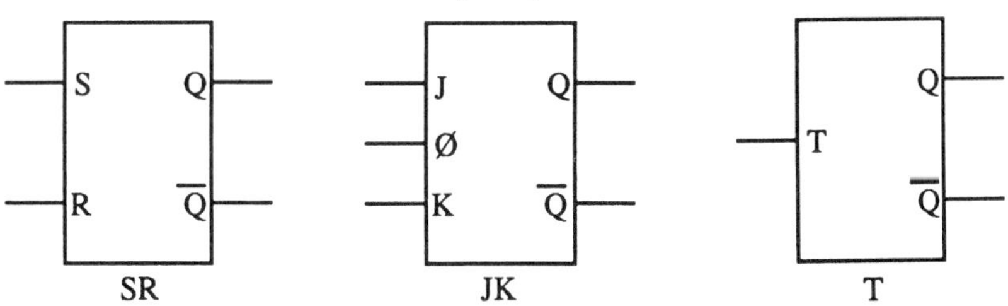

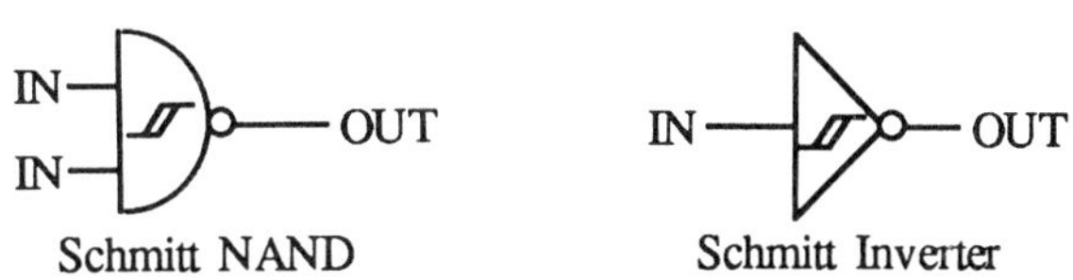

Miscellaneous Symbols

Ac receptacles/plugs

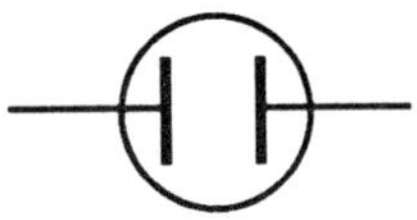

Non-polarized

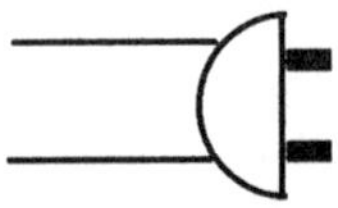

Polarized

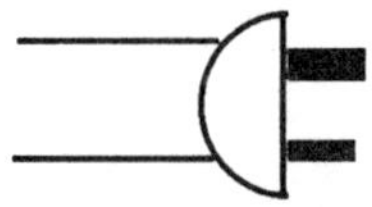

Ac voltage sources

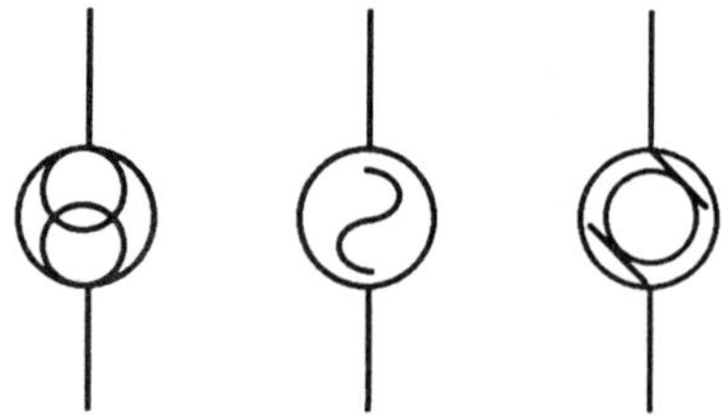

Antennas

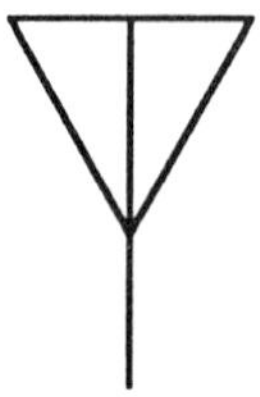

General

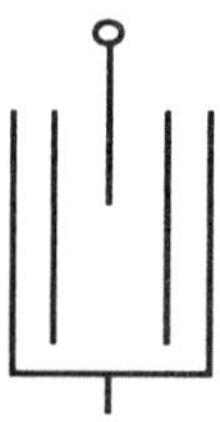

Telescoping

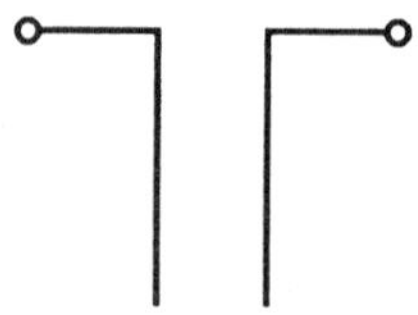

Dipole

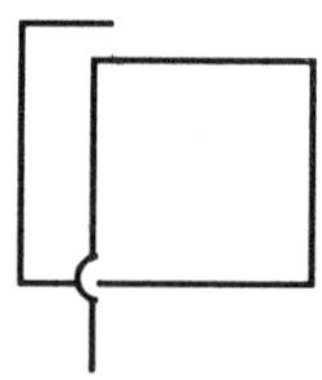

General

Batteries

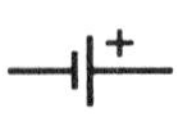

Single cell

Multiple cell

Circuit breakers

Fuses

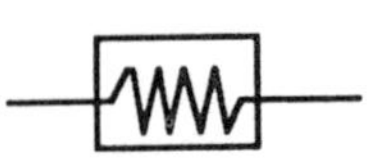

Headphones

Standard double

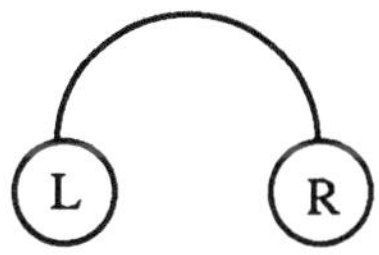

Stereo/binaural

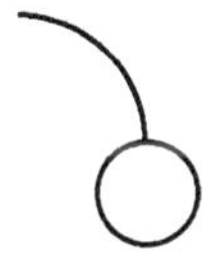
Half

Lamps

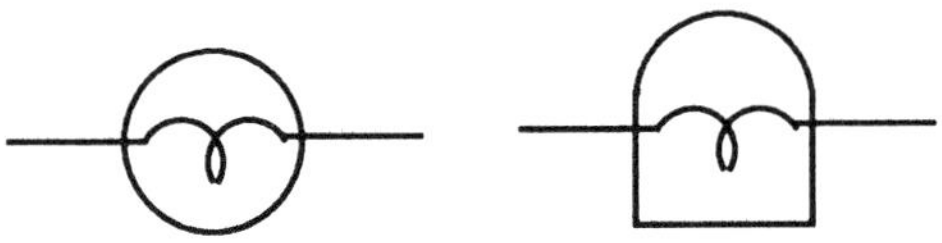
Incandescent

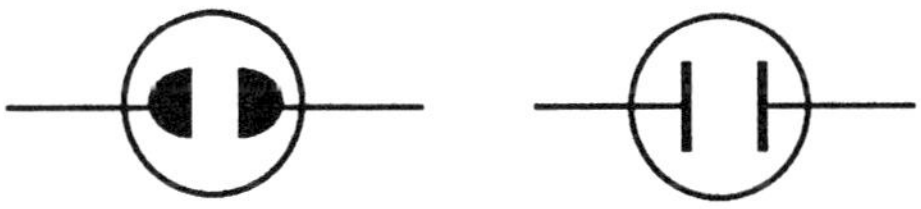
Neon

Meter movement

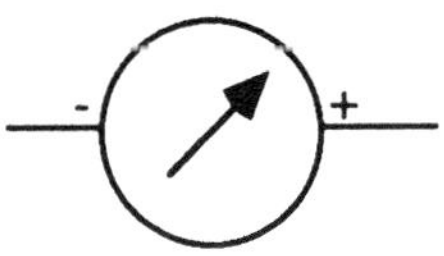

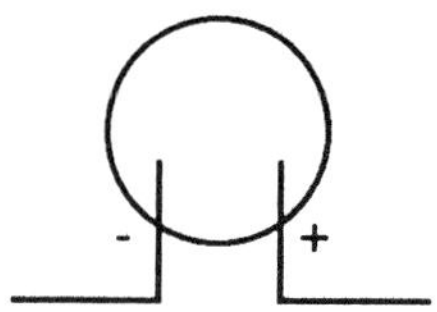

Microphones

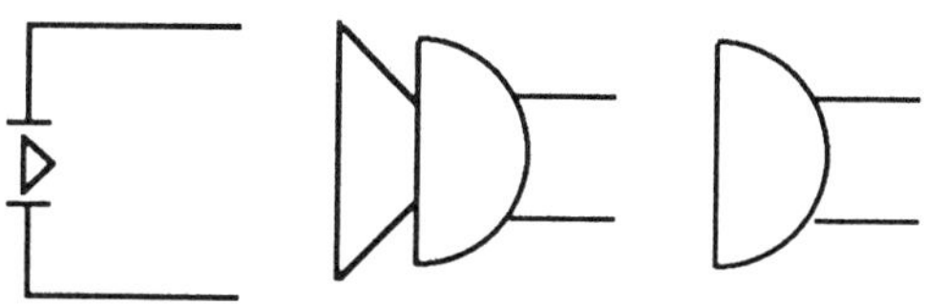

Motors

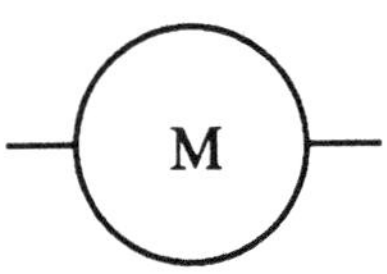

Basic

Single-phase
(more coils denote
additional phases)

Solar cells

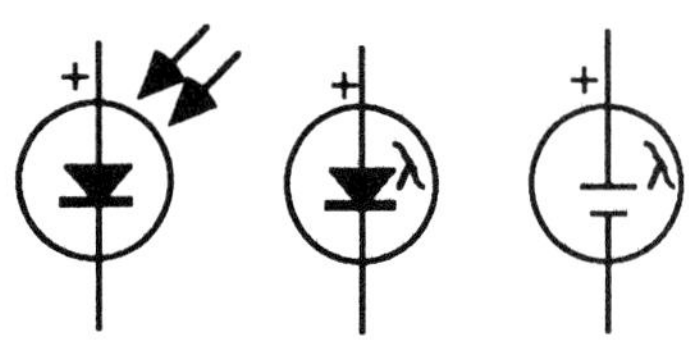

Solenoids

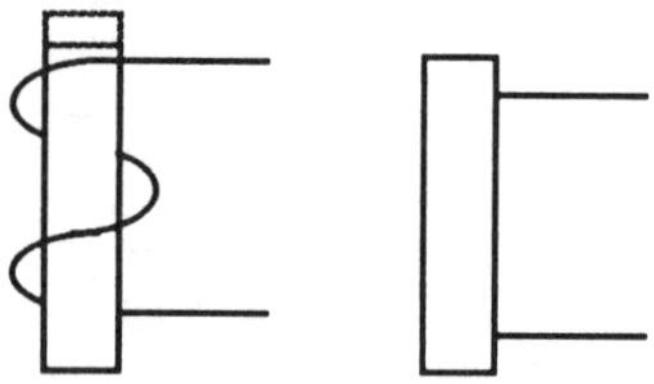

Speakers

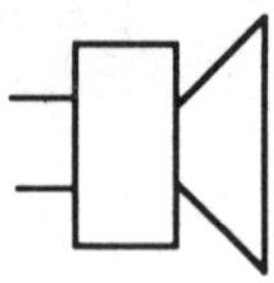

General

Electrostatic

Dynamic

Switches

SPST SPDT DPST DPDT

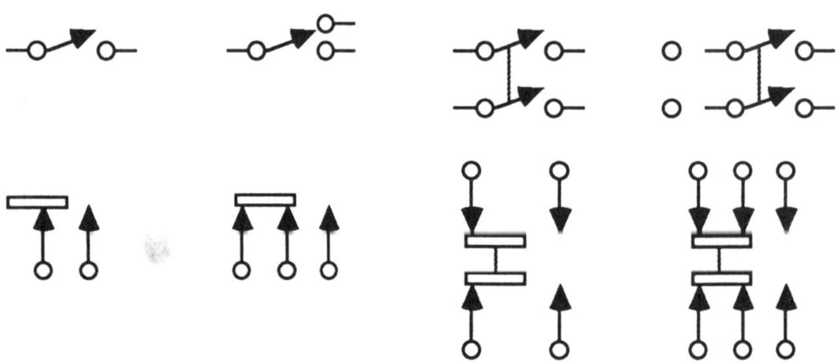

Pushbutton

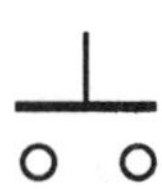

Normally open (NO)

Normally closed (NC)

8

All about batteries

MOST HOBBY ELECTRONIC CIRCUITS RUN ON BATTERY POWER. UNDERSTANDING how they work, and the differences in types, will help you choose the correct battery for the job.

This chapter details the most common batteries you'll encounter in your hobby electronics work. It also describes methods of improving power distribution in your circuits and how to properly apply fuse protection to prevent overload and short-circuit burnout.

Battery Types

There are six main types of commercially available batteries: zinc, alkaline, nickel-cadmium, lead-acid, gelled electrolyte, and lithium (other battery types exist, of course, but they are either not regularly available to hobbyists or they are far too expensive for homebrew projects). These types come in a variety of shapes, sizes, and configurations.

Zinc

Zinc batteries are the staple of the battery industry and are often referred to simply as ''flashlight'' cells. The chemical makeup of zinc batteries comes in two forms: carbon zinc and zinc chloride. Carbon zinc, or ''regular duty,'' batteries die out the quickest and are unsuited for most electronics applications. Zinc chloride, or ''heavy duty,'' batteries provide a little more power than regular carbon zinc cells and last 25 to 50 percent longer. Both carbon zinc and zinc chloride batteries can be rejuvenated a few times after being drained by using a battery charger.

Alkaline

Alkaline cells use a special alkaline manganese dioxide formula that lasts up to 800 percent longer than carbon zinc batteries. The actual increase in life expectancy ranges from about 300 percent to 800 percent, depending on the application. Unlike zinc batteries, however, alkaline batteries cannot be rejuvenated by recharging. When the battery is dead, throw it away.

Nickel-cadmium

When you think "rechargeable battery," you undoubtedly think nickel-cadmium, or "ni-cad." Ni-cads aren't the only battery specifically engineered to be recharged, but they are among the least expensive and easiest to get. With the exception of driving large motors, ni-cads are ideal for most all electronics applications.

The cells are available in all standard sizes, plus special-purpose "sub" sizes for use in sealed battery packs (as in rechargeable handheld vacuum cleaners, photoflash equipment, and so forth). Most subsized batteries have solder tabs on them, so you can directly attach wires to the battery instead of placing the cells in a battery holder. After use, the ni-cads are placed in a recharger and recharged for 8 to 16 hours.

A new, higher capacity ni-cad battery recently became available that offers two to three times the service life of regular ni-cads and can be recharged more quickly (generally under 5 or 6 hours). More importantly, the high-capacity cells provide considerably more power and are ideally suited for more demanding electronics work. Of course, cost is higher.

Lead-acid

The battery in your car is a lead-acid battery. Its makeup is not much more than lead plates crammed in a container that's filled with an acid-based electrolyte. These brutes pack a wallop and have an admirable between-charge life. When the battery goes dead, recharge it, just like a ni-cad.

Not all lead-acid batteries are as big as the one in your car. You can also get—new or surplus— 6-volt lead-acid batteries that are about the size of a small radio. The battery is sealed so the acid doesn't spill out (most automotive batteries are now sealed as well). The sealing isn't complete but provides a blowhole to allow gases to escape during recharging.

Lead-acid batteries typically come in self-contained packs. Six-volt packs are the most common, but you can also get 12- and 24-volt packs. The packs are actually made by combining several smaller cells. The cells are wired together to provide the rated voltage of the entire pack. Each cell typically provides 1.5 volts, so three cells are required to make a 6-volt pack. You can, if you want, take the pack apart, unsolder the cells, and use them separately.

Motorcycle batteries make good power cells for heavy-duty projects. They are easy to get, compact, and relatively lightweight. The batteries come in various amp-hour capacities (discussed later in this chapter), so you can choose the best one for your application. Motorcycle batteries are somewhat pricey, however. Car batteries can also be used, but they usually weigh in excess of 25 pounds.

Gelled electrolyte

Gelled-electrolyte batteries are also rechargeable and provide high current for a reasonable period of time, making them perfect for many heavy-duty applications. Gelled-electrolyte batteries are typically available only in 6- or 12-volt packs. Some specially designed cells (available through surplus) come with different voltage ratings.

Lithium

Lithium batteries are used whenever long shelf life is desired. Their most common application is as low-voltage backup power for computer memory. The IBM PC/AT and clones use a lithium battery to provide backup power to maintain the current time and date, as well as current system parameters. The battery delivers about 5 volts continuously for a period of several years. Lithium batteries are also used in many digital wrist watches, hearing aids, and other miniature electronics where long battery life is desired.

Lithium batteries are capable of delivering a hefty amount of current (for their size) for a long period of time, after which they die quickly. Figure 8-1 shows the discharge curve for a lithium battery as compared to some of the other types we have discussed.

When not in use and stored in a cool, dry place, lithium batteries maintain over 95 percent of their rated voltage and current for several years. The price you pay for the long life of lithium batteries is high cost: lithium batteries in AA or C size can cost upwards of $10 to $20 each. Cost is somewhat less in the surplus market.

While certain types of lithium batteries can be recharged, the type most often used in electronics are not. Recharging a nonrechargeable lithium battery is not recommended.

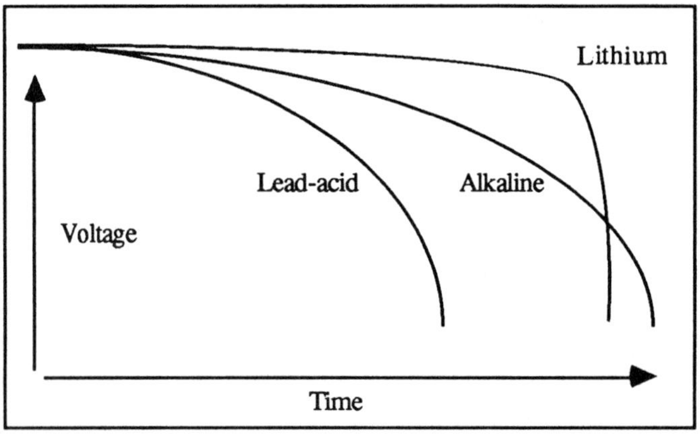

Fig. 8-1. Typical battery discharge curves for alkaline, lithium, and lead-acid batteries.

Ni-Cad Pluses and Minuses

Ni-cad batteries are among the most popular batteries for hobby electronics because they are relatively inexpensive, can be recharged quickly and economically, and have a reasonably long service life. Despite their numerous advantages, however, ni-cads have a few odd peculiarities you'll want to consider when designing your portable projects.

The most annoying problem is what's termed "memory effect," which is most often noted in the standard variety, rather than the high-output kind. Ni-cads tend to remember the rate at which they have been discharged and recharged in the past and tend to follow this curve in subsequent uses. If you always use the battery to 50 percent of its capacity and then recharge it, the cell will develop a memory where it will last as little as half as long as it should before giving out.

Recent developments in ni-cad design have gone a long way to reduce the affliction of memory effect, but it can still be troublesome if you always recharge your batteries before depleting them.

Memory effect is altered in two ways:

- Short the battery until it's dead. Recharge it as usual. Some batteries may be permanently damaged with this technique.
- Use the battery in a low-current circuit, like a flashlight, until it is dead. Recharge the battery as usual. You must repeat this process a few times until the memory effect is gone.

The best way to combat memory effect is to avoid it in the first place. Always fully discharge ni-cad batteries before charging them. If you don't have a flashlight handy, build yourself a discharge circuit using a battery holder and a flashlight bulb, as shown in Fig. 8-2. The bulb acts as a "discharge" indicator.

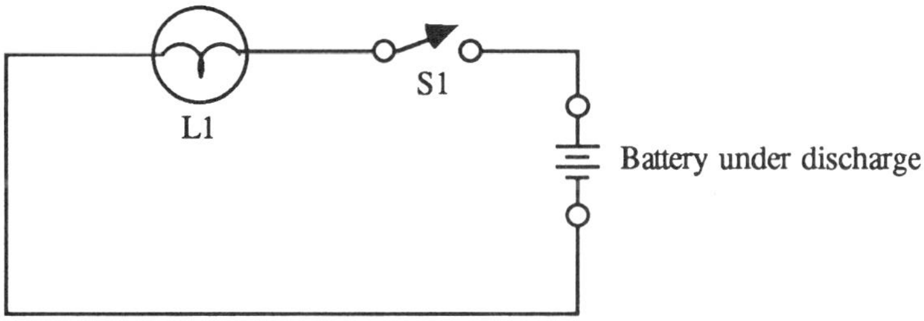

L1 3-12 Vdc lamp (as required by battery)
S1 SPST toggle

Fig. 8-2. A flashlight bulb battery "discharger."

When it goes out completely, the batteries are fully discharged.

The other disadvantage is that the polarity of ni-cads can change—positive becomes negative and vice versa—under certain circumstances. Polarity reversal is somewhat common if the battery is left discharged too long, or if it is discharged below 75 or 80 percent capacity. Excessive discharging can occur if one or more cells in a battery pack wears out. The adjacent cells must work overtime to compensate, discharging themsleves too fast and too far.

You can test for polarity reversal by hooking the battery to a volt-ohmmeter (remove it from the pack if necessary). If you get a negative reading when the leads are connected properly, the polarity of the cell is reversed.

Polarity reversal can sometimes be corrected by fully charging the battery (connecting it in the recharger in reverse), then shorting it out. Repeat the process a couple of times, if necessary. You have about a 50-50 chance that the battery will survive this. The alternative is to throw the battery out, so you actually stand to lose very little.

About Battery Rates

Batteries carry all sorts of specifications. The two most important specifications are voltage and amp-hour current.

Voltage

The voltage rating of a battery is fairly straightforward. If the cell is rated for 1.5 volts, it puts out 1.5 volts, give or take. That "give or take" is more important than you might think because few batteries actually deliver their rated voltage throughout their life span. Most rechargeable batteries are recharged 20 to 30 percent higher than their specified rating. For example, the 12-volt battery in your car, a type of lead-acid battery, is charged to about 13.8 volts.

Standard zinc and alkaline flashlight batteries are rated at 1.5 volts per cell. Assuming a well-made battery in the first place, the voltage might actually be

1.65 volts when the cell is fresh or 1.2 volts when the battery is considered "dead." The circuit or motor you are powering with the battery must be able to operate sufficiently throughout this range. A listing of "fresh" and "dead" voltage ratings for various batteries is shown in Table 8-1.

When experimenting with your projects, keep a volt-ohmmeter handy and periodically test the output of the batteries. Perform the test while the battery is in use. The test results might be erroneous if the battery is not tested under load.

Table 8-1. Battery Voltage Levels

	Primary (alkaline, etc.)	Lead-acid	Ni-cad
Nominal	1.5	2	1.2
Fully charge	1.65	2.15	1.4
Discharge	1.25	1.75	0.95

Amp-hour current

The amp-hour (Ah) current is the amount of power, in amps or milliamps, the battery can deliver over a specified period of time. What exactly does the term *amp-hour* mean? Basically, the battery will be able to provide the rated current for one hour before failing. But there's more to it than this. The 5 Ah rating is actually taken at a 10- or 20-hour discharge interval. That is, the battery is used for 10 or 20 hours at a low or medium discharge rate. After the specified time, the battery is tested to see how much juice it has left. The rating of the battery is then calculated taking the difference between the discharge rate and the reserve power and multiplying it by the number of hours under test.

What all this means is that it's an unusual battery that provides the stated amps in the one hour period. The battery is much more likely to fail after 30 or 45 minutes of heavy-duty use and won't be able to supply the specified current for more than about 15 or 20 minutes. Discharging at or above the amp-hour rating could actually cause damage to the battery. This is especially true of ni-cad cells.

Recharge rate

Batteries must be recharged slowly, over a 12- to 24-hour period. The battery can't take on too much current without breaking down and destroying itself, so the current from the battery charger must be kept at a safe level.

A good rule of thumb to follow when recharging any battery is to limit the recharging level to one-tenth the amp-hour rating of the cell. For a 5 Ah battery, then, a safe recharge level is 500 milliamps. Current limit is extremely important when recharging standard capacity ni-cads, which can be permanently damaged by charging too quickly. Lead-acid, gelled-electrolyte, and many high-capacity ni-cad batteries can take an occasional ''fast-charge,'' a quickie at 25 to 50 percent of the rated amp-hour capacity of the battery.

The recharge period, the number of hours the battery is recharged, varies depending on the type of cell. The recharge rate is usually a function of the discharge rate. You can't recharge a battery any faster than it is discharged. Figure 8-3 shows a typical discharge/recharge curve for a lead-acid or gelled-electrolyte battery.

Most manufacturers specify the recharge time for their batteries. If so, follow it. If no recharge time is specified, assume a 3:1 or 4:1 discharge/recharge ratio, and place the battery under charge for that period of time. Continue experimenting until you find an optimum recharge interval.

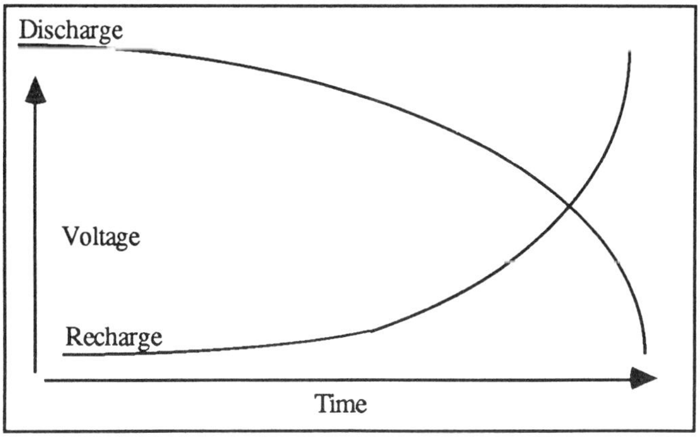

Fig. 8-3. Discharge/recharge curve.

Increasing battery ratings

You can obtain higher voltages and current by connecting several cells together, as shown in Fig. 8-4.

- To increase voltage, connect the batteries in series. The resultant voltage is the sum of the voltage outputs of all the cells combined.
- To increase current, connect the batteries in parallel. The resultant current is the sum of the current capacities of all the cells combined.

Take note that when connecting cells together, all the cells might not discharge or recharge at the same rate. This is particularly true if you combine two half-used batteries with two new ones. The new ones will do the lion's share of the work and won't last as long. Therefore, always replace or recharge all the cells at once.

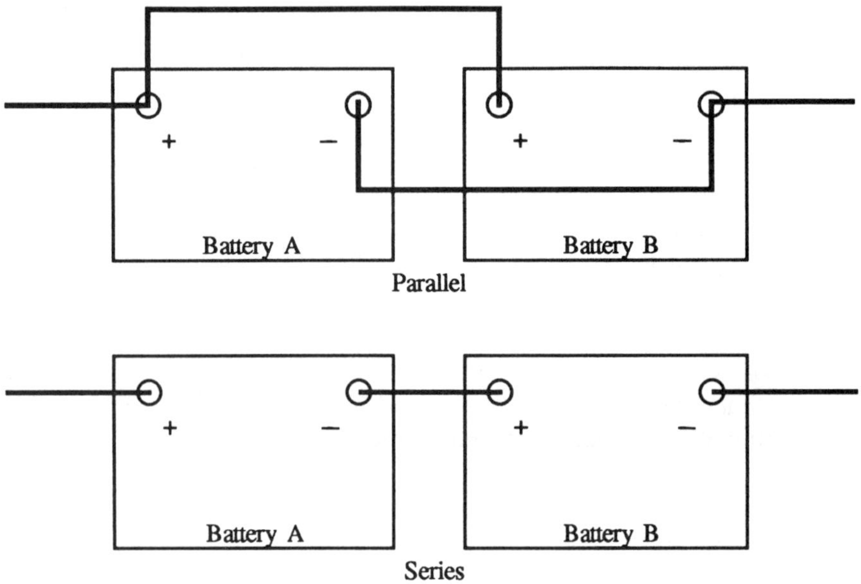

Fig. 8-4. How to connect batteries in parallel and series.

Ni-cad, lead-acid, and gelled-electrolyte batteries should be periodically recharged, whether they need it or not. Once every two or three months is sufficient for lead-acid and gelled electrolyte; once every four to five months is adequate for ni-cad cells. Always observe polarity when recharging batteries. Inserting the cells backward in the recharger will destroy the batteries and possibly damage the recharger.

Battery Care

Batteries are rather sturdy little creatures, but you should follow some simple guidelines when using them. Your batteries will last much longer, so you'll save yourself some money.

- Store new batteries in the refrigerator (not the freezer). Put them in a plastic bag to prevent contamination from possible leakage. Warm up the batteries for several hours before using them.
- Avoid using or storing batteries in temperatures above 75 or 80 degrees F. The life of the battery will be severely shortened otherwise. Using a battery above 100 to 125 degrees causes rapid deterioration.
- Unless you're repairing a misbehaving ni-cad, avoid shorting out the terminals of the battery. Besides possibly igniting fumes exhausted by the battery, the sudden and intense current output shortens the life of the cell.

- Keep rechargeable batteries charged. Make a note when the battery was last charged.
- Fully discharge ni-cads before charging them again to prevent memory effect.
- Given the right circumstances, all batteries can leak, even the "sealed" variety. When not in use, keep batteries in a safe place where leaked electrolyte cannot cause damage. Remove batteries from their holder when not in use.

Providing Power

Now that you know about batteries, you can start using them in your electronics designs. The most simple and straightforward arrangement is to use a commercial-made battery holder. Holders are available that contain from two to eight AA, C, or D batteries, such as that shown in Fig. 8-5.

Wiring in the holders connects the batteries in series, so a four-cell holder puts out 6 Vdc (1.5 times 4). You attach the leads of the holder (red for positive

Fig. 8-5. A sampler of battery holders.

and black for ground or negative) to the main power supply rail in your circuit. You'd follow a similar procedure if using a gel-cell or lead-acid battery.

Fuse protection

Flashlight batteries don't deliver extraordinary current, so fuse protection is not required on the most basic projects. Gelled-electrolyte, lead-acid, and high-capacity ni-cad batteries can deliver a most shocking amount of current, enough that if the leads of the battery accidentally touch each other or there is a short in the circuit, the wires could melt and a fire could erupt.

Fuse protection helps eliminate the calamity of a short circuit or power overload. As illustrated in Fig. 8-6, connect the fuse in line with the positive rail of the battery, as near to the battery as possible. You can purchase fuse holders that connect directly to the wire, or mount it on a panel or printed circuit board.

Choosing the right value of fuse can be a little tricky, but not impossible. It does require that you know how much current the circuit draws from the battery during operation. You can roughly figure the fuse you need by calculating the current draw from the load. Other active components in the circuit also draw power, so consider these as well. After calculations are complete, tack on 20 to 25 percent overhead.

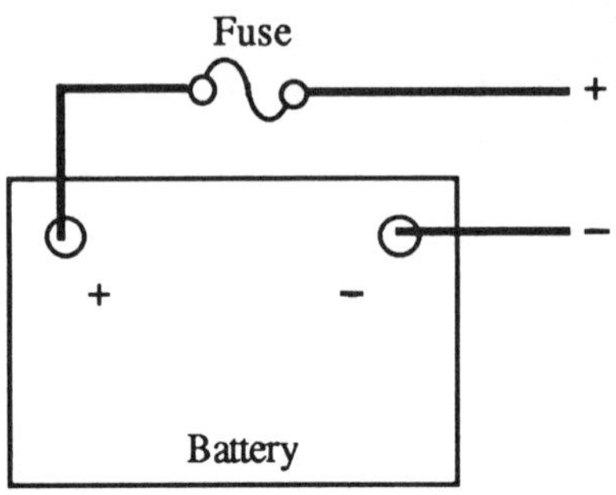

Fig. 8-6. Adding a fuse to the positive battery rail.

A more certain method is to directly measure current consumption. Place a meter in line with the negative supply rail, as shown in Fig. 8-7. Dial the meter to read current and apply power. The meter will indicate the exact current draw of the circuit.

Note that this method should only be used on circuits that draw less than the current capacity of the meter, normally 1 or 2 amps (up to 10 amps with a current shunt). If the current draw exceeds the maximum rated capacity of the meter, the internal fuse in the meter could blow, or the meter could be damaged. Chances are, unless the circuit powers a motor or other high-current load, current draw will be a reasonable 100 to 1000 milliamps. Once you know the load of the circuit, add 20 to 25 percent for protection, and select the appropriate fuse.

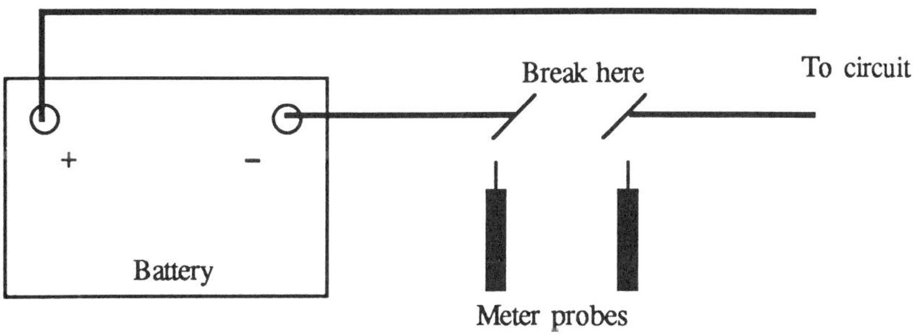

Fig. 8-7. How to measure battery current with meter. Be sure the meter is rated for the current you are testing, or you could blow a fuse in the VOM.

Fuses don't come in every conceivable size or value. Standard values for fuses are:

1/4 amp	4 amps
1/2 amp	5 amps
3/4 amp	6 amps
1 amp	7 amps
1 1/2 amps	8 amps
2 amps	10 amps
2 1/2 amps	15 amps
3 amps	20 amps

If your circuit consumes 1,750 milliamps (1 3/4 amps), use at least a 2-amp fuse. In most cases, a 3- or 4-amp fuse can also be used even though the amperage is more than twice the current consumption of the circuit. Many circuits require extra current when they are turned on. This extra surge could cause the fuse to blow unless it is "slow-blow" type.

For the sake of standardization, choose the regular 1 1/4-by-1/4-inch-diameter bus fuses for your circuits. You'll have an easier job finding fuse holders for them, and you have a greater selection of values. Even with a standard fuse size, there is not much to choose from past 8 amps other than 10, 15, and 20 amps, but this is acceptable for most applications.

Power distribution

Figure 8-8 shows how to distribute the power from a single battery source to many separate circuit boards. The individual regulators, which are optional, provide power for one or two large boards or a half dozen or so smaller ones.

Voltage regulators are a familiar sight on ac-operated circuits, but they are not as common in circuits that run on battery power. However, if your circuit requires voltage regulation, there's nothing stopping you from using them

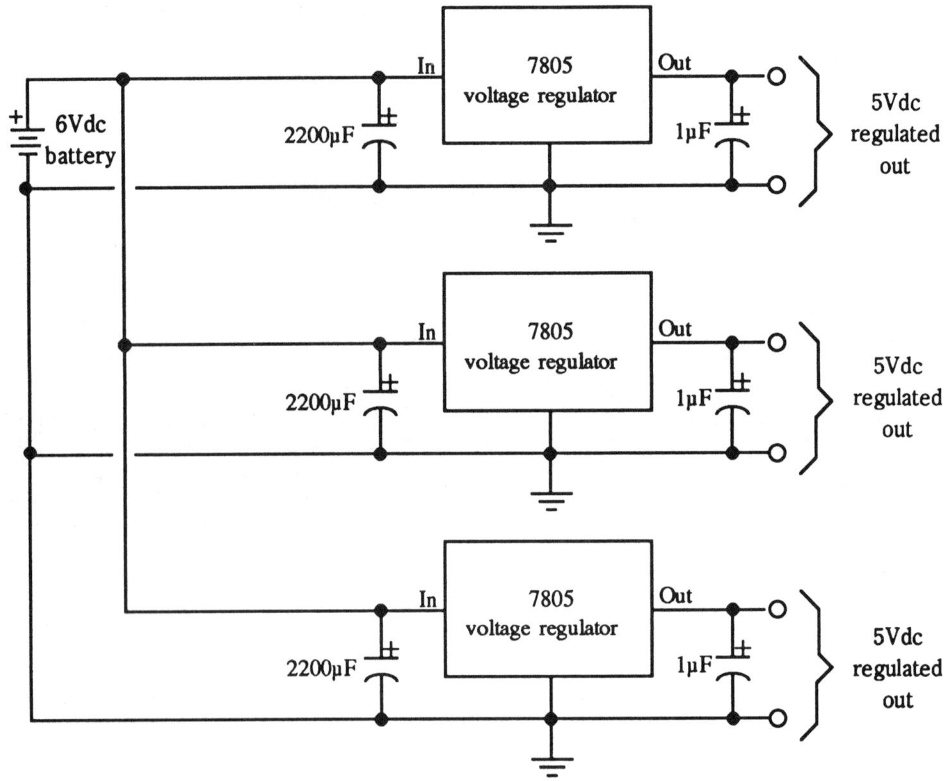

Fig. 8-8. Multiple regulation or single battery supply.

because voltage regulators simply regulate dc volts. Note that all regulators require a slightly higher operating voltage than the desired regulated output. For example, if you need 5.0 volts dc, you need to supply 6.2 to 7 volts to the regulator.

Note the bypass capacitors strategically positioned on the subsystem boards. These help prevent glitches from passing between one circuit board to another through the power supply lines. The bypass capacitors are not strictly required, but they are highly recommended, especially if your circuits use many TTL integrated circuits.

9

Component identification

COMPONENTS ARE THE THINGS THAT MAKE YOUR ELECTRONIC PROJECTS WORK. The average hobby project contains about three dozen components of varying types, sizes, and styles including resistors, capacitors, diodes, integrated circuits, and light-emitting diodes.

This chapter summarizes the components commonly found in hobby electronic projects and discusses their various forms and varieties. You'll learn how to identify parts by sight, determine their value, and test for basic component well-being.

Fixed Resistors

A fixed resistor supplies a predetermined resistance to a circuit. The standard unit of value of a resistor is the ohm, shown schematically as Ω. The higher the ohm value, the more resistance the component provides to the circuit.

The value on most fixed resistors is identified with color coding, shown in Fig. 9-1. The color coding starts near the edge of the resistor and comprises four, five, and sometimes six bands of different colors. Most off-the-shelf resistors for hobby projects use standard four-band color coding.

With the four-band system, bands one, two, and three indicate the value of the resistor. Band four indicates the tolerance of the resistor, which is usually 5 or 10 percent.

High-precision resistors often employ a five-band color coding system.

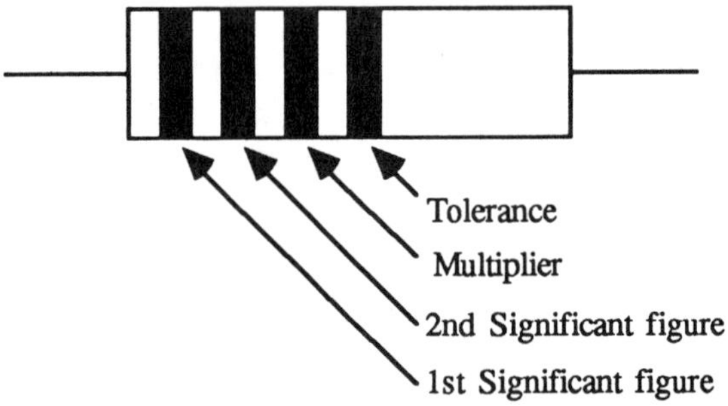

Fig. 9-1. The four-band resistor color coding.

Bands one through four indicate the value of the resistor, and the fifth band, as shown in Fig. 9-2, indicates the tolerance of the resistor, typically 1 percent.

Table 9-1 shows the meaning of the color codes. Here's how to use the table. Assume the resistor uses the four-band color coding symbol and is colored yellow-violet-red-silver. The first two bands indicate the "significant digits" of the value of the resistor. Table 9-1 shows that yellow is 4 and violet is 7, so the significant digits of the resistor are 47. The third band indicates the multiplier, in this case 100. Multiply 47 by 100, and you get 4700 ohms. Values over a thousand are most often expressed with the letter k, which stands for *kilo* (thousand), so the resistor is said to have a value of 4.7 k (for 4.7 thousand) or 4.7 kΩ (4.7 thousand ohms).

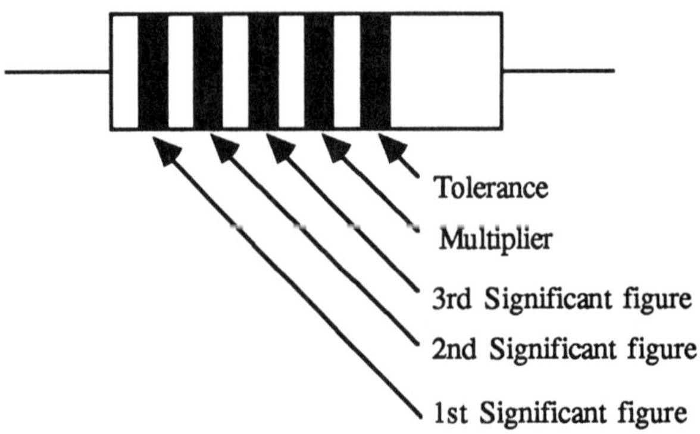

Fig. 9-2. The five-band resistor color coding.

The fourth band of the resistor indicates its tolerance, in this case 10 percent. That means that the resistor can vary in tolerance 10 percent higher or lower than its indicated value. If the resistor has an indicated value of 4.7 k, with a 10-percent tolerance, the actual value could be anywhere between 4,230 to 5,170 ohms.

Table 9-1. Resistor Color Coding

Color	1st Digit	2nd Digit	Multiplier	Tolerance
Black	0	0	×1	± 20%
Brown	1	1	×10	± 1%
Red	2	2	×100	± 2%
Orange	3	3	×1,000	± 3%
Yellow	4	4	×10,000	± 4%
Green	5	5	×100,000	
Blue	6	6	×1,000,000	
Violet	7	7	×10,000,000	
Gray	8	8	×100,000,000	
White	9	9	—	
Gold			×0.1	± 5%
Silver			×0.01	± 10%
None				± 0%

In the five-band color-coding system, the first three bands are used for the significant digits. Suppose the resistor is marked blue-gray-red-orange-brown. Here's how you'd decode the colors:

Significant digits			Multiplier	Tolerance
blue	gray	red	orange	brown
6	8	2	1000	1%

Multiply 682 by 1000 (or simply add three zeros after 682), and the result is 682,000 ohms, or 682 k. The tolerance is only one percent, or ±6,820 ohms.

If you are not sure of the resistance of a particular resistor, use a volt-ohm-meter to check it. Position the test leads on either end of the resistor, as shown in Fig. 9-3. If the meter is not autoranging, start at a high range and work down. Be sure that you don't touch the test leads or the leads of the resistor because if you do, you'll add the natural resistance of your own body into the reading.

Resistors are also rated by their wattage. The wattage of a resistor indicates the amount of power it can safely dissipate. Resistors used in high-load applications, like motor or lamp control, require higher wattages than those used for low-current uses.

The majority of resistors you'll use for hobby electronics will be rated at 1/4 or even 1/8 of a watt. The wattage of a resistor is not marked on the body of the component; rather you must infer the wattage by the size of the resistor. Figure 9-4 shows several resistors with common wattages of 1/8, 1/4, 1/2, and 1 watt.

Resistors up to about 2 watts are most commonly the carbon composition type. That is, the resistor is composed of carbon granules cemented around a

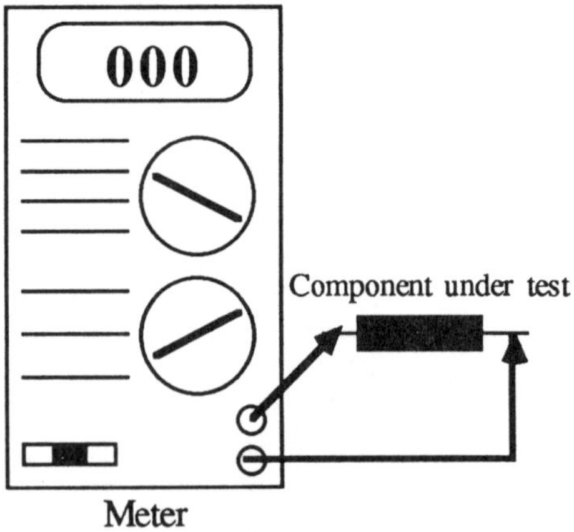

Fig. 9-3. Testing a resistor with a VOM.

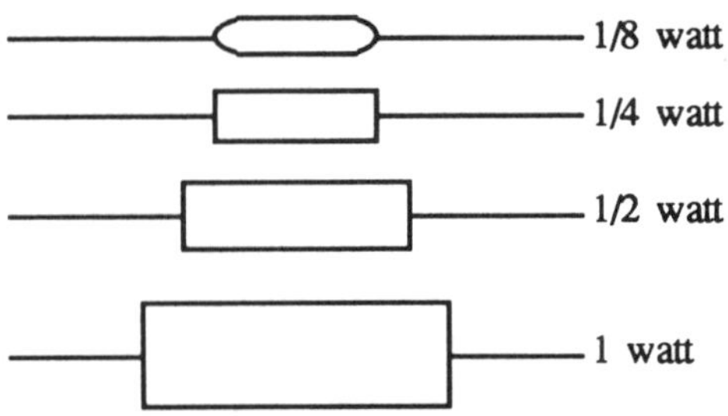

Fig. 9-4. Most of the resistors you'll work with will be rated at 1/8, 1/4, 1/2, and 1 watt. The respective sizes of each are shown here.

pair of electrical leads. The density of the carbon largely determines the resistance—the more carbon in the component, the lower the resistance.

Higher tolerance resistors most often use a metal-film composition. The construction of metal-film resistors differs between manufacturers, but many consist of a very thin film of sputtered or evaporated metal (such as nickel-chromium) bonded to a ceramic substrate. Metal-film resistors are commonly under 5 or 6 watts.

Resistors with power ratings of 3 watts or more are typically the wire-wound type. These are composed of a high-tensile wire wound around a nonconducting core. The diameter of the wire and its length determine resistance.

High-wattage resistors take many forms, some of which are shown in Fig. 9-5. Resistors over 5 watts are commonly encased in silica or other waterproof, flameproof coating and have a rectangular—rather than cylindrical—shape. Even higher wattage resistors might include their own metal heatsink. The fins help draw heat away from the resistor.

Variable Resistors

Variable resistors, more commonly known as potentiometers, let you ''dial in'' a specific resistance. The actual range of resistance is determined by the upward

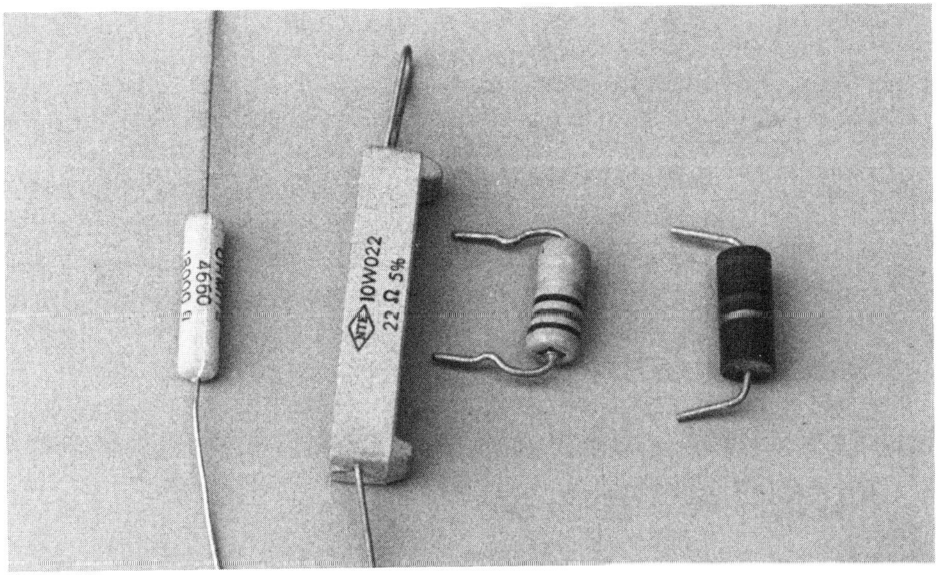

Fig. 9-5. A sampler of high-wattage resistors.

value of the potentiometer. Potentiometers are thus marked with this upward value such as 10 k, 50 k, 100 k, 1 M, and so forth. For example, a 50 k potentiometer lets you dial in any resistance from 0 ohms to 50,000 ohms. Note that the range is approximate.

Potentiometers are either the dial or slide type, as shown in Fig. 9-6. The dial type is the most familiar, used in such applications as television volume controls or electric blanket thermostat controls. The rotation of the dial is nearly 360 degrees, depending on the potentiometer used. In one extreme, the resistance through the potentiometer (or "pot") is zero; in the other extreme, the resistance is the maximum value of the component.

Both dial and slide pots use similar construction. A thin conductive material is wound or bonded to a ceramic form. The ends of this material form two of the

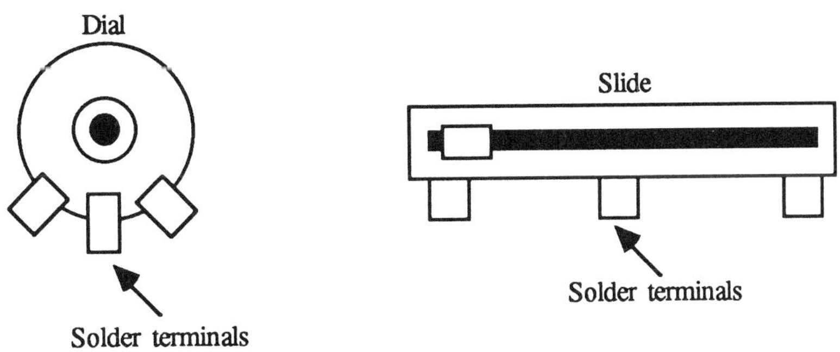

Fig. 9-6. Dial- and slide-type potentiometers.

connections to the potentiometer. A metal tab on the knob (or slide) forms the third connection. This tab is referred to as the *wiper*. As you turn the dial or move the slide, the wiper travels along the conductive material. If you apply a volt-ohmmeter between one end of the conductive material and the wiper, you'll note a change in resistance as you move the dial or slide.

Note, as shown in Fig. 9-7, that you can connect the meter to either end of the conductive material. This merely reverses the operation of the pot. For example, with the meter applied to points 1 and 3, turning the dial clockwise *increases* resistance. Conversely, with the meter applied to points 2 and 3, turning the dial clockwise *decreases* resistance.

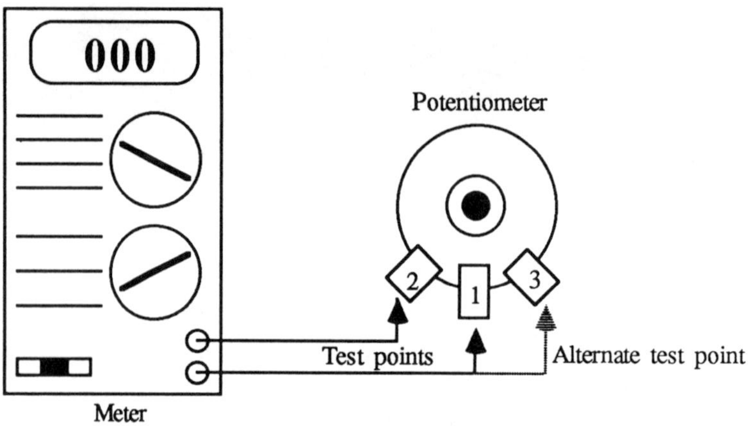

Fig. 9-7. How to test a meter: connect the probes to the outside terminals and to the center wiper.

The material used to form the conductive surface of the pot can take many forms, including *cermet* (a combination of ceramic, glass, and precious metals), carbon, wire, and conductive plastic. In most cases, you won't be able to choose the conductive material that is used in the potentiometers for your projects.

The upward or maximum value of a potentiometer is most often stamped somewhere on the component. If the pot lacks markings, you'll need to use a meter (applying the test probe leads to points 1 and 2) to determine the value of the component.

As with fixed resistors, potentiometers are rated by their wattage. Most pots are rated at 0.5 watts; heavier duty potentiometers are available with a rating of up to about 3 watts. Some pots have even higher wattage ratings, but these are reserved for special applications. For example, L-pads, used in audio applications, are high-power variable resistors with typical wattage ratings of 25, 50, or even 100 watts. The resistance of L-pads is generally low, however, on the order of a few hundred ohms.

Some projects require precision potentiometers, referred to as *multiturn pots* and *trimmers*. Instead of turning the dial one complete rotation to change the resistance from, say, 0 to 10,000 ohms, a multiturn pot requires you to rotate the knob 3, 5, 10, or even 15 times to span the same range. Most are designed for mounting directly on the printed circuit board and require a screwdriver or plastic tool for adjustment.

Finally, there is one additional aspect of potentiometers that determines their application. As you turn the knob or move the slide, the resistance of a pot changes either linearly or semilogarithmically. In most potentiometers, the change is linear so that as you move the control, the output of the pot changes in direct proportion. For example, if you rotate the pot halfway, its resistance changes by 50 percent. Such a potentiometer is called *linear taper*.

In a semilogarithmic or audio-taper potentiometer, the output of the component changes logarithmically as you turn the knob. This type of potentiometer is most often used to control audio levels, which follow a logarithmic scale. You can tell the difference between an audio-taper and linear-taper potentiometer by connecting a meter between the wiper and one end of the component (such as points 1 and 3, as shown in Fig. 9-7). Dial the knob halfway. If the measurement is 50 percent of the total resistance of the pot, it is linear taper. If the measurement is about 15 or 85 percent of the total resistance of the pot, it is audio taper.

Fixed Capacitors

After resistors, capacitors are the second most common component found in the average electronic project. Capacitors serve many purposes, including removing traces of alternating current ripple in a power supply, delaying the action of some portion of the circuit or removing harmful glitches. In all cases, these applications rely on the ability of the capacitor to hold an electrical charge for a predetermined period of time.

Capacitors come in many more sizes, shapes, and varieties than resistors, though only a small handful are truly common. However, most capacitors are composed of the same basic stuff: a pair of conductive elements separated by an insulating dielectric (see Fig. 9-8). This dielectric can be composed of many materials including air (in the case of a variable capacitor, as detailed in the next section), paper, epoxy, plastic, or even oil.

In most capacitors, there are actually many layers of conducting elements and dielectric. The additional layers help increase the capacitance of the component. The layers are electrically connected to provide two leads coming from the body of the capacitor.

Capacitors are rated by their capacitance, in farads, and by the breakdown voltage of their dielectric. The farad is a rather large unit of measurement, so the bulk of capacitors available today are rated in microfarads (a millionth of a farad).

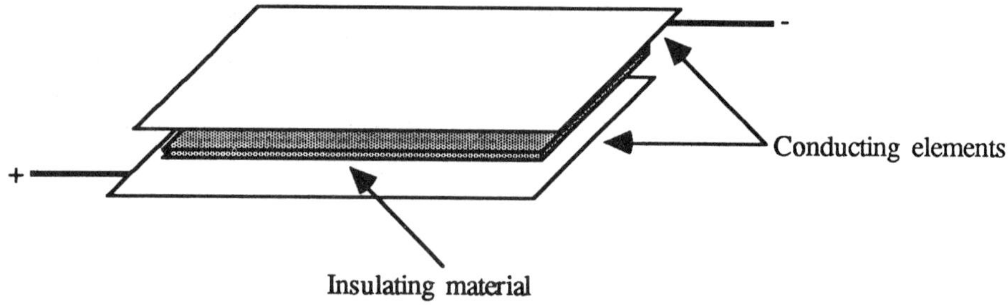

Fig. 9-8. Typical construction of capacitor.

An even smaller rating is the picofarad, or a millionth of a millionth of a microfarad. For example, a 10-microfarad capacitor is 10 millionths of a farad. A 1-microfarad capacitor is 1 millionth of a farad. Microfarad is most often shortened using the greek "micro" character, to μF, as in 10 μF. The picofarad is simply shortened to pF.

As an aside, only until recently have farad-range capacitors been possible. Using older fashioned construction techniques, a 1-farad capacitor might measure a city block and weigh thousands of tons. But by using a unique formula of granulated carbon, 1-farad capacitors can now be built that are not much larger than the volume control knob on your TV set. These capacitors are routinely used as low-power electrical backup for computer memories, clock radios, and other electric devices that need to retain a small charge for extended periods of time when power is removed.

The voltage rating is the highest voltage the capacitor can withstand before there is damage to the dielectric layers in the component. High voltages cause nasty sparks—the higher the voltage, the longer and more powerful the spark. If a capacitor is not designed to withstand high voltages, a spark can develop within the capacitor that can punch through the dielectric material, rendering the component useless (generally shorted out).

Capacitors are largely classified by the dielectric material they use. The most common dielectric materials are aluminum electrolytic, tantalum electrolytic, ceramic, mica, polypropylene, polyester (or Mylar), paper, and polystyrene. When you select a capacitor for a particular job, you must generally also indicate the type, such as ceramic, mica, or Mylar.

The dielectric material partly determines its useful application. The larger electrolytic capacitors, which use an aluminum electrolyte, are suited for such chores as power supply filtering, where large values are needed. Table 9-2 lists the most common capacitor types, their typical value range, typical tolerance (as a percentage of their stated value), and common applications.

Capacitors come in a variety of shapes, as shown in Fig. 9-9. A common shape for aluminum electrolytic and paper capacitors is cylindrical. Tantalum electrolytic, ceramic, mica, and polystyrene capacitors are more bulbous shaped

Table 9-2. Capacitor Characteristics

Type	Range	Application
Ceramic	1 pF to 2.2 μF	Filtering, bypass
Mica	1 pF to 1 μF	Timing, oscillator, precision circuits
Metalized polycarbonate	To 100 μF	Dc blocking, power supply filtering, coupling
Polyester	0.001 to 100 μF	Same as polycarbonate
Polystyrene	To 10 μF	Timing, tuning circuits
Paper foil	To 100 μF	General purpose
Tantalum	0.001 to 1000 μF	Bypass, coupling, dc blocking
Aluminum electrolytic	To 220,000 μF	Filtering, coupling, bypass

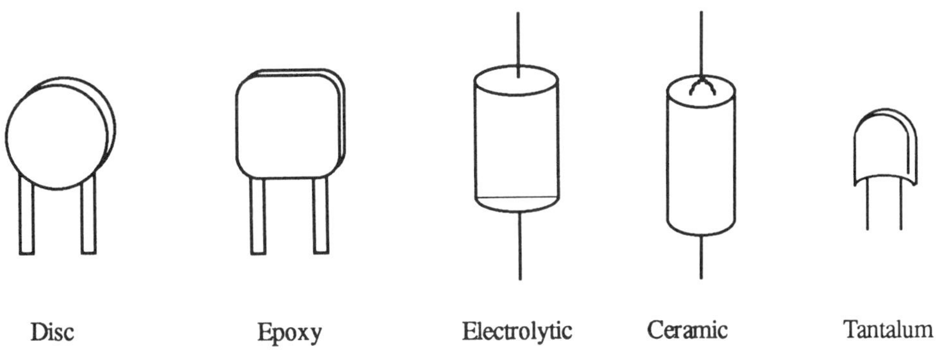

Fig. 9-9. Popular capacitor styles.

because they are typically dipped into an epoxy or plastic bath to form the outside skin. Not all capacitors of any particular type (such as mica, Mylar, paper, etc.) are manufactured the same way, so you can't always tell the construction of the component just by looking at it.

The values for many capacitors are printed directly on the component. This is especially true with the larger aluminum electrolytic, where the large size of the capacitor provides ample room to print the capacitance and voltage.

Smaller capacitors, such as 0.1 and 0.01 μF mica disc capacitors, as shown in Fig. 9-10, use a common three-digit marking system to denote capacitance and tolerance. The numbering system is easy to use, if you remember it's based on picofarads, not microfarads. A number such as 104 means 10, followed by four zeros, as in "100,000," or 100,000 picofarads. Values over 1,000 picofarads are most often stated in microfarads. To make the conversion, move the decimal point to the left six spaces. Result: 0.1 μF. Table 9-3 is a handy reference guide showing common capacitor markings using this numbering system.

Fig. 9-10. Many disc capacitors are marked with three digits, as shown here. (The text tells you how to decode the markings.)

Table 9-3.
Capacitor Value Reference

Marking	Value (μF)
01 to 99	pF
101	0.0001
102	0.001
103	0.01
104	0.1
221	0.00022
222	0.0022
223	0.022
224	0.22
331	0.00033
332	0.0033
333	0.033
334	0.33
471	0.00047
472	0.0047
473	0.047
474	0.47

Note that values under 1000 picofarads do not use this numbering system. Rather, the actual value, in picofarads, is listed, such as 10 (for 10 pF).

Another, less common numbering system uses numbers and letters, as in "4R7." The letter R indicates the position of the decimal point, as in 4.7. This numbering system doesn't indicate the units of measure, however, which might be microfarads, picofarads, or nanofarads (a billionth of a farad).

Though it is becoming rare, capacitors are also sometimes color coded. The colors could be painted in bands, as with resistors, or painted as dots on the flat surface of the capacitor (see Fig. 9-11). In both cases, capacitors use the same color coding as resistors.

You can test capacitance with a capacitor meter, or a volt-ohmmeter with a capacitance input. Most meters require that you plug the capacitor directly into the test instrument because capacitance can vary with long lead lengths.

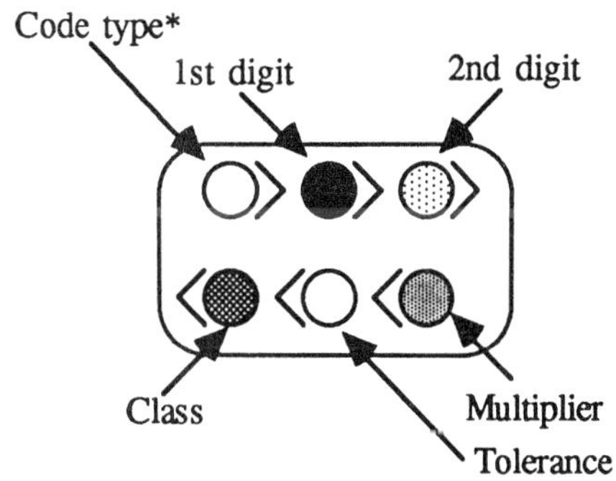

Fig. 9-11. Older-style color-dot-coded capacitors use resistor color coding.

White = EIA code
Black = Mil
Silver = AWS (paper)

The tolerance of the capacitor is most often indicated by a single letter code, sometimes placed by itself on the body of the capacitor or placed after the three-digit mark, such as "104Z." The letter Z donates a tolerance of +80 percent and −20 percent. That means the capacitor, which is rated at 0.1 μF, might be as much as 80 percent higher or 20 percent lower. Table 9-4 shows the meaning of other common code letters used to indicate capacitor tolerance.

Table 9-4.
Capacitor Tolerance Markings

Code	Tolerance
B	± 0.1 pF
C	± 0.25 pF
D	± 0.5 pF
F	± 1%
G	± 2%
J	± 5%
K	± 10%
M	± 20%
Z	+ 80%, −20%

The tolerance of a capacitor changes with temperature. Temperature stability is commonly designated using a three-character code, such as NPO, which stands for negative/positive zero. A capacitor with the NPO designation is going to be fairly tolerant of temperature changes.

More and more capacitor manufacturers are adopting the EIA marking system for temperature tolerance, shown in Table 9-5. The three characters in the

Table 9-5. EIA Capacitor Codes

Letter Symbol	Low Temp. Requirement
Z	+10 deg. C
Y	−30 deg. C
X	−55 deg. C

Number Symbol	High Temp. Requirement
2	+ 45 deg. C
4	+ 65 deg. C
5	+ 85 deg. C
6	+105 deg. C
7	+125 deg. C

Letter Symbol	Max. Capacitance Change over Temp. Rating
A	± 1.0%
B	± 1.5%
C	± 2.2%
D	± 3.3%
E	± 4.7%
F	± 7.5%
P	± 10.0%
R	± 15.0%
S	± 22.0%
T	± 22%-33%
U	± 22%-56%
V	± 22%-82%

EXAMPLE: Y5P = ± 10% variation of temperature over range of −30 deg. C to +85 deg. C.

mark indicate the temperature tolerance and maximum variation within the stated temperature range. For example, a capacitor marked Y5P has the following characteristics:

- −30 C low-temperature requirement
- +85 C high-temperature requirement
- ±10.0 percent variance in capacitance over the −30 to +85 range

The maximum dielectric breakdown voltage is not always stated on the body of a capacitor, but if it is, it is almost always indicated by the actual voltage, such as "35" or "35V." Sometimes, the letters WV follow the voltage rating, which indicates *working voltage* (really, the maximum dielectric breakdown voltage) of the capacitor. You should not use the capacitor with voltages exceeding this rating. The voltage rating is most often stated in dc, but some capacitors (especially electrolytics) indicate ac voltage or ac and dc voltages.

One final mark you will find almost exclusively on larger tantalum and aluminum electrolytic is a polarity symbol, typically a minus (–) sign. The polarity symbol indicates the positive and/or negative lead of a capacitor. For example, as shown in Fig. 9-12, the minus sign and arrow point to the negative lead of the aluminum electrolytic capacitor.

Note that only the larger value (1 μF and up) capacitors are generally polarized, and these are most often limited just to the electrolytic types (nonpolarized electrolytic capacitors are available too, commonly used in stereo speaker systems). The smaller capacitors, such as mica, ceramic, and Mylar capacitors are not polarized, so they don't have a polarity mark.

If a capacitor is polarized, it is extremely important to install it in the circuit following the proper orientation. If you reverse the leads to the capacitor—connect the + side to the ground rail, for example—the capacitor could be ruined. Other components in the circuit could also be damaged.

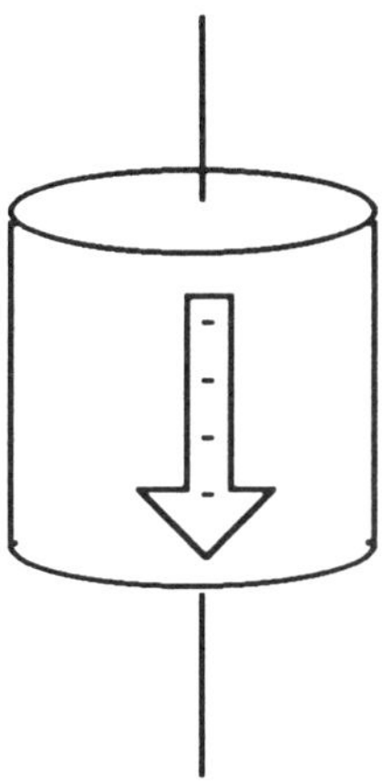

Fig. 9-12. Always observe the polarity on polarized electrolytic capacitors.

Variable Capacitors

Variable capacitors are similar to variable resistors in that they allow you to adjust capacitance to suit your needs. Unlike potentiometers, however, variable capacitors operate on a drastically reduced range of values, and they seldom provide "zero" capacitance.

The most common type of variable capacitor is the air dielectric type, like those found in the tuning control of an AM radio. As you dial the tuning knob, you move one set of plates within another. Air, the dielectric, separates the plates so they don't touch.

Smaller variable capacitors are sometimes used as "trimmers" for adjusting the capacitance within a narrow band. You will often find trimmers in radio receivers and transmitters and in circuits that use quartz crystals for an accurate reference signal. The value of such trimmers is typically in the 5 to 30 pF range.

Inductors

Inductors are often used to increase or decrease a circuit's sensitivity to a particular frequency. Inductors are commonly used in radio receivers and transmitters, but they can also be found in any other type of circuit as well. For example, inductors are often used in the more elaborate power supplies as a means to help reduce the 60 Hz component of the incoming power line.

Inductors are sometimes known by other names including *choke* and *coil*. These names can often be used interchangeably, but they don't always mean the same thing. For example, a coil is most often used to resonate at a certain frequency, and a choke is used to affect a group or range of frequencies. I use the word *inductor* throughout the remainder of this section as a term that encompasses all subvarieties.

Inductors can be both fixed or variable. In either, a slender wire is wound around an insulating core. The number of turns of the wire and its diameter determine the value of the inductor. The core can be air, a ferrite compound, or any of a number of other materials (though air and ferrite are most common). Small-value inductors look much like low-wattage resistors and even have similar color coding marks—in fact, the color coding for inductors is the same as it is for resistors. Larger inductors can take many forms.

Variable inductors are often used in RF circuits where some measure of adjustment is required. A variable inductor is composed of a wire wound around a hollow core. A ferrite slug is placed in the core and is moved back and forth to change the inductance.

The value of an inductor is stated in henrys, or more commonly, millihenrys (thousandths of a henry) and microhenrys (millionths of a henry). As mentioned earlier, the value of the inductor is usually marked using the same color-coding technique employed for resistors. The values of larger and variable inductors are often printed directly on the component.

Inductors are also rated by several other factors including Q and maximum current. However, these ratings are seldom printed on the component. When purchasing new inductors, this information is typically provided in a table or chart in the ordering catalog.

Diodes

The diode is the most simple form of semiconductor. Diodes are available in two basic flavors: germanium and silicon. These types indicate the material used to manufacture the active junction within the diode. Silicon diodes are by far the most common.

Diodes are used in a variety of applications, and there are numerous subtypes. Here is a list of the most common:

- *Rectifier*—The "average" diode rectifies ac current to provide dc only.
- *Zener*—Limits voltage to a predetermined value. Often used for low-cost voltage regulation.
- *Light emitting*—Emits infrared of visible light when current is applied.
- *Silicon-controlled rectifier* (SCR)—A type of low-power switch used to control ac or dc currents.
- *Bridge rectifier*—A collection of four diodes strung together in sequence; used for rectifying an incoming ac current so that all waves of the current are present.

Other types of diodes include the diac, triac, bilateral switch, light-activated SCR, and several other variations. A sampling of diodes is shown in Fig. 9-13.

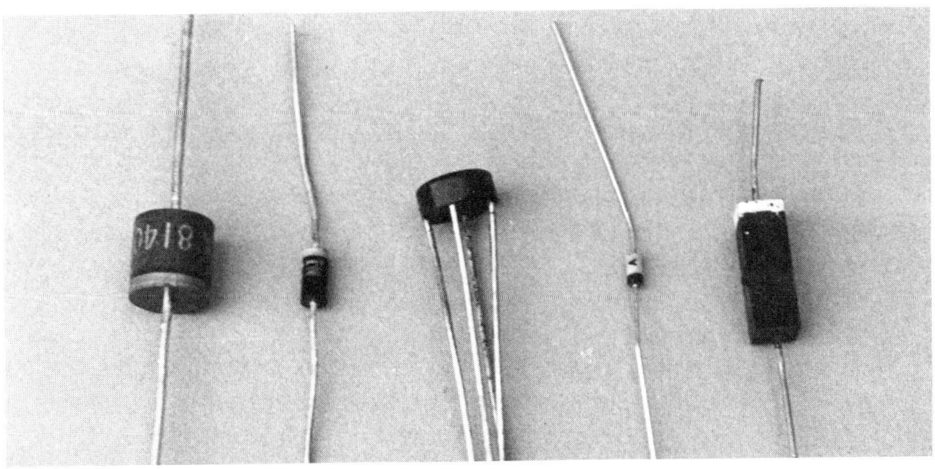

Fig. 9-13. A sampler of diodes. The larger diodes are rated for higher voltages, higher amps, or both.

Diodes carry two important ratings: ***peak inverse voltage*** (PIV) and ***current***. The PIV rating roughly indicates the maximum working voltage for the diode. For example, if the diode is rated at 100 volts, it should not be used in a circuit that passes more than 100 volts.

Similarly, the current rating is the maximum amount of current the diode can withstand. Assuming a diode is rated for 3 amps, it cannot safely conduct more than 3 amps without overheating and failing.

Rectifier diodes rated to about 3 to 5 amps are generally encased in a black or gray epoxy case and are designed for directly mounting on a printed circuit board. Higher current diodes, such as 20, 30, or 40 amps, are commonly contained inside a metal housing. The metal housing includes a mounting stud so that the diode can be securely affixed to a heatsink. A few diodes use the same packaging as transistors (described in the next section).

All diodes have positive and negative terminals. The positive terminal is the *anode*, and the negative terminal is the *cathode*. You can readily identify the cathode end of a diode by looking for a red or black stripe near one of the leads. Figure 9-14 shows a diode with a stripe at the cathode end and how the stripe corresponds with the heavy line in the schematic symbol for the diode.

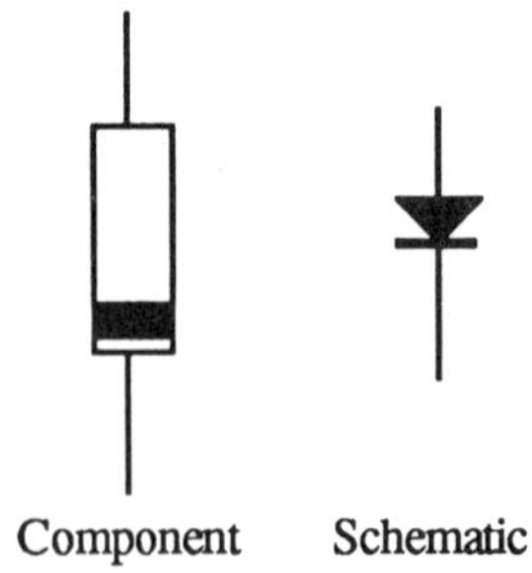

Fig. 9-14. Diode showing cathode stripe, and schematic symbol.

All semiconductors emit light when an electric current is applied to them. This light is generally very dim and only in the infrared region of the electromagnetic spectrum. The light-emitting diode (LED) is a special type of semiconductor that is expressly designed to emit copious amounts of light. Most LEDs are engineered to produce red, yellow, or green light, but special-purpose types are designed to emit infrared and blue light.

The LED has a PIV rating of about 100 to 150 volts with a maximum current rating of under 40 milliamps. Most LEDs are used in low-power dc circuits and are powered with 12 volts or less. Even though this voltage is far below the PIV rating of the LED, the component can still be ruthlessly damaged if it is exposed to currents exceeding 40 or 50 mA. A resistor is used to limit the current to the LED.

You cannot test the PIV and current rating of a diode just by looking at it. However, you can test it if the diode is functional and you have a volt-ohmmeter that has a diode-check setting. To test the component, dial the meter to the diode-check setting. Apply the test leads of the meter to either side of the diode, as shown in Fig. 9-15. Observe proper polarity: the red test lead to the anode of the diode, and the black test lead to the cathode. Remember to avoid touching the test leads with your fingers. Observe the reading. Now reverse the leads and test again.

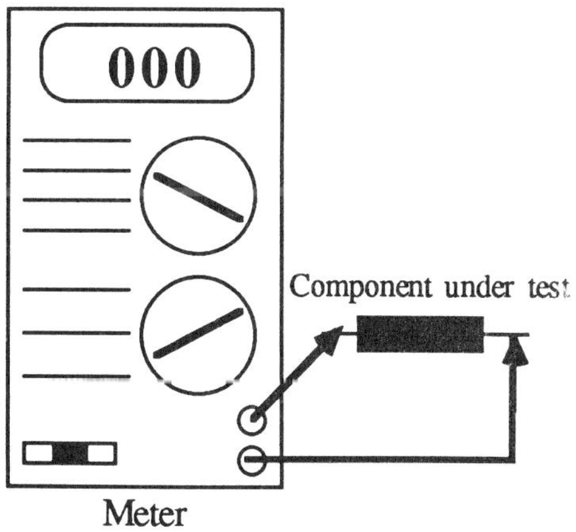

Fig. 9-15. How to test a diode.

Here's how to interpret the results (note: this test is applicable to most diode types except LEDS, which you can most often test visually):

Display Value 1st	2nd	Condition/Device Type
About 0.4	Overrange	Good/germanium
About 0.6	Overrange	Good/silicon
Overrange	Overrange	Bad—open
Small or zero	Small or zero	Bad—short

Transistors

Transistors were designed as an alternative to the old vacuum tube and are used in similar applications, either to amplify a signal or to switch a signal on and off. At last count there were several *thousand* different transistors currently available from over two dozen manufacturers.

Besides amplifying or switching a current, transistors are used for two broad categories:

- *Signal*—Used with relatively low-current circuits like radios, telephones, and most other hobby electronics projects.
- *Power*—Used with high-current circuits like motor drivers and power supplies.

You can most often tell the difference between the two merely by size. The signal transistor is rarely larger than a pea and uses slender wire leads. The power transistor uses a large metal case to help dissipate heat and uses heavy spokelike leads. Figure 9-16 shows a comparison between the two types.

Transistors are identified by a unique code, such as 2N2222 or MPS6519. You need to refer to a data book to ascertain the characteristics and ratings of the

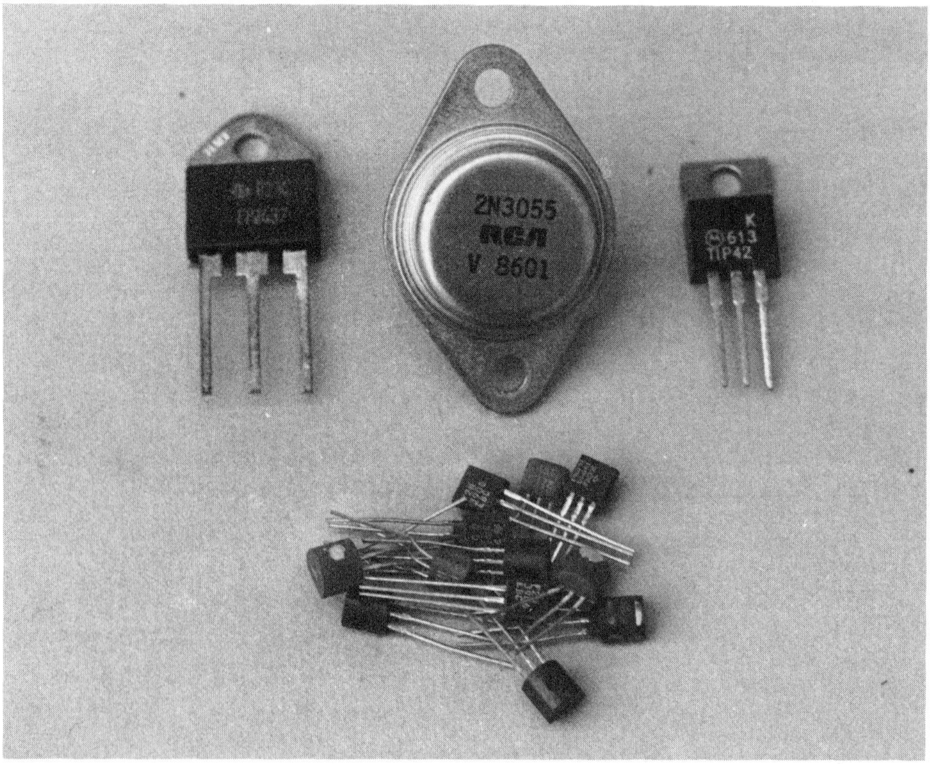

Fig. 9-16. Sampler of signal and power transistors. Some signal transistors come in metal TO-5 canisters, which are slightly larger than the plastic ones shown here.

particular transistor you are interested in. Transistors are rated by a number of criteria, which are far too extensive for the scope of this book. These ratings include collector-to-base voltage, collector-to-emitter voltage, maximum collector current, maximum device dissipation, and maximum operating frequency. None of these ratings are printed directly on the transistor.

Other markings might be visible on the transistor, however, such as date codes (to show the date of manufacture), a manufacturer code, or a manufacturer's in-house classification or catalog number.

Signal transistors are available in either plastic or metal cases. The plastic kind is suitable for most uses, but some precision applications require the metal variety. Transistors that use metal cases (or "cans") are less susceptible to stray radio frequency interference. They also dissipate heat more readily.

Power transistors always come in metal cases, though a portion of the case (the back or sides) might be made of plastic.

Transistors have three or four wire leads. The leads in the typical three-lead transistor are the base, emitter, and collector, shown in Fig. 9-17. A few transistors, most notably the field-effect transistor (or FET) have a fourth lead. This is for grounding the case to the chassis of the circuit.

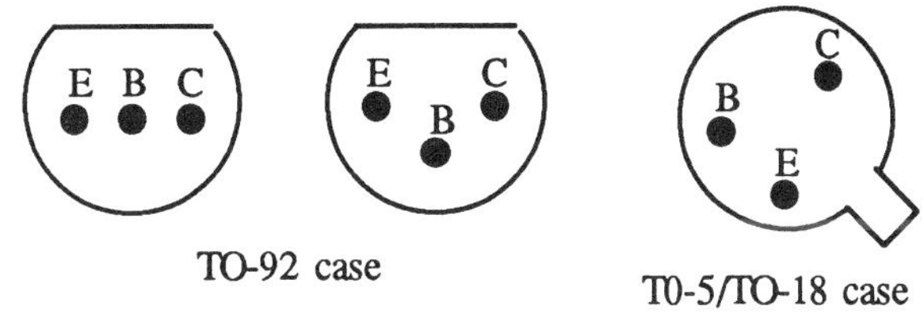

Fig. 9-17. Underside package view of a three-lead transistor.

Transistors can be either npn or pnp devices. This nomenclature refers to the sandwiching of semiconductor materials inside the device. You can't tell the difference between an npn and pnp transistor just by looking at it. However, the difference is indicated in catalog specifications sheet, as well as schematically, as shown in Fig. 9-18.

Some semiconductor devices look and act like transistors and they are called transistors, but they actually use a different technology. For example, the MOSFET (for metal-oxide semiconductor field-effect transistor) is occasionally used in circuits that demand high current and high tolerance. MOSFET transistors don't use the standard base-emitter-collector connections; rather, they are called the gate, drain, and source, as depicted in Fig. 9-19. Note, too, that the schematic diagram for the MOSFET is different than for the standard transistor.

Fig. 9-18. Npn/pnp transistor schematic symbols.

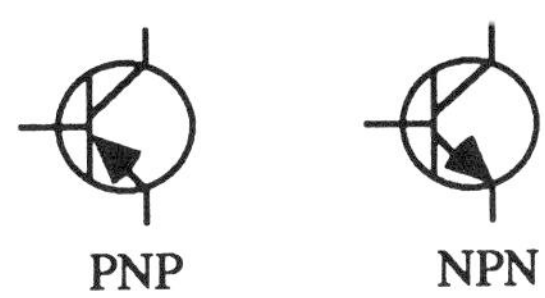

Fig. 9-19. Schematic representation of a FET.

Integrated Circuits

The integrated circuit forms the backbone of the electronics revolution. The typical integrated circuit comprises many transistors, diodes, resistors, and even capacitors. As its name implies, the integrated circuit, or IC, is a discrete and wholly functioning circuit in its own right. ICs are the building blocks of larger circuits. You merely string them together to form just about any project you can envision.

Integrated circuits are most often enclosed in dual in-line-pin (DIP) packages, as shown in Fig. 9-20. The illustration shows several sizes of DIP ICs,

Fig. 9-20. Assortment of ICs.

from those with 8 to 40 pins. The most common are 8-, 14-, and 16-pin. The IC can be either soldered directly into the circuit board or mounted in a socket.

As with transistors, ICs are identified by a unique code such as 7400 or 4017. This code indicates the type of device. Specifications and parameters about the IC can then be looked up in a reference book. The code is printed on the top or "back" of the IC, as depicted in Fig. 9-21, so it can be readily identified.

Many ICs also contain other written information including manufacturer catalog number and date code. Do not confuse the date code or catalog number with the code used to identify the device.

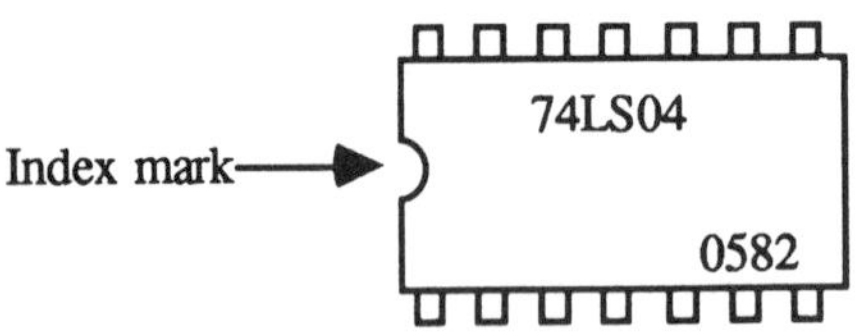

Fig. 9-21. The index mark on an IC denotes pin 1.

10

Common IC pinout diagrams

HOW OFTEN HAVE YOU BEEN BUILDING A PROJECT AND THEN HAD TO STOP because you forgot the pinout assignments for an integrated circuit? You can always hunt through a data book, but the organization of the book might not help you easily find the pinout diagram you nccd.

This chapter provides dozens of pinout diagrams for the most common IC and transistor components including diagrams for popular TTL, CMOS, and linear ICs.

TTL Integrated Circuits

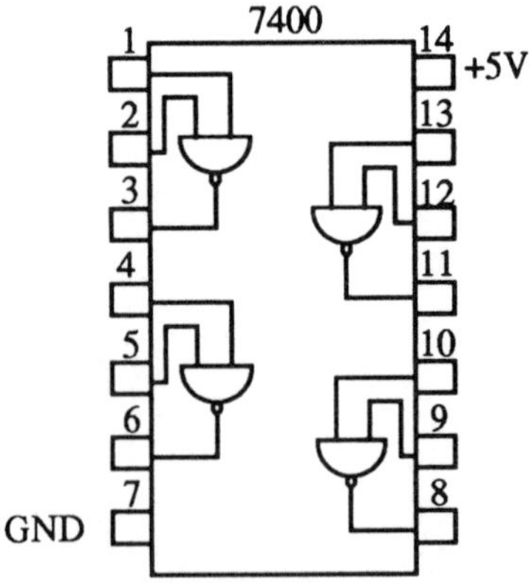

(Same as 7403)
(Same as 7437)

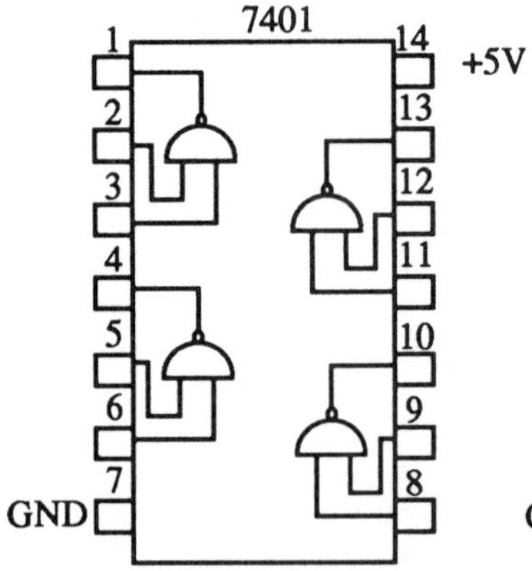

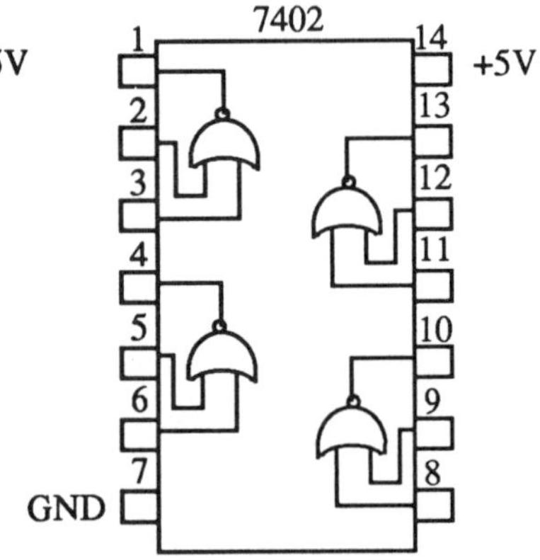

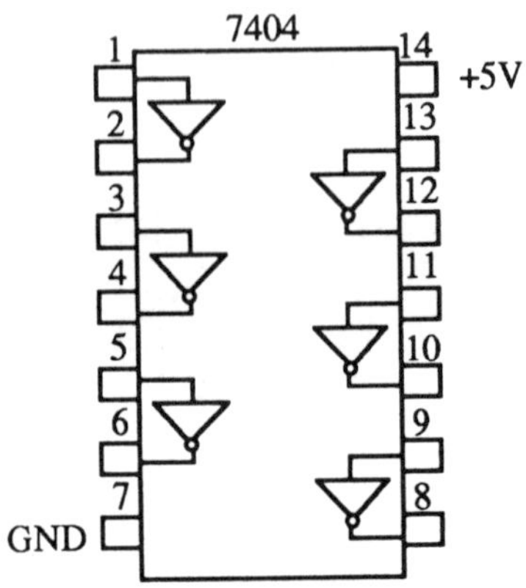

(Same as 7405)
(Same as 7406)
(Same as 7416)

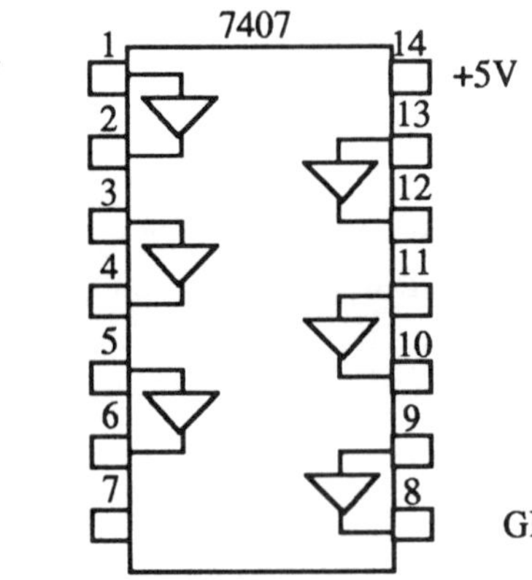

(Same as 7417)

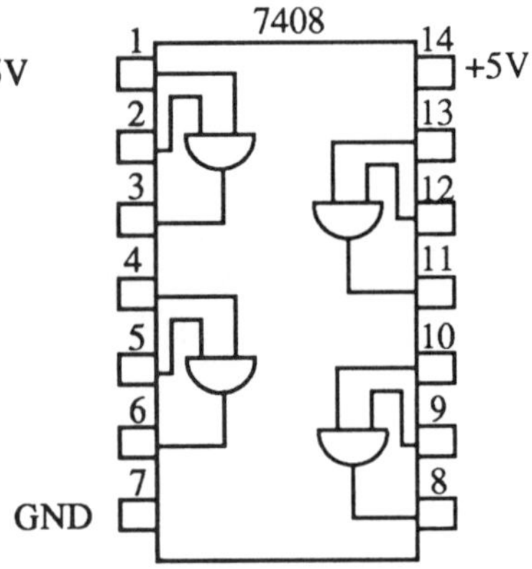

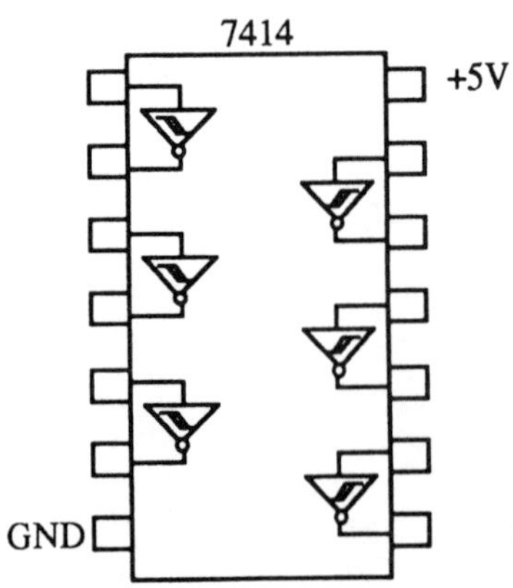

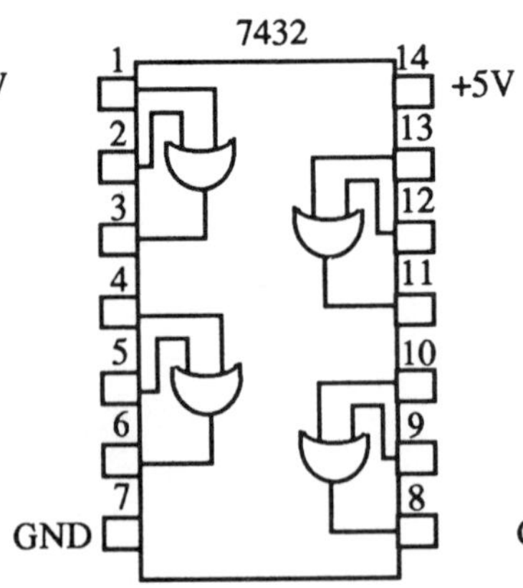

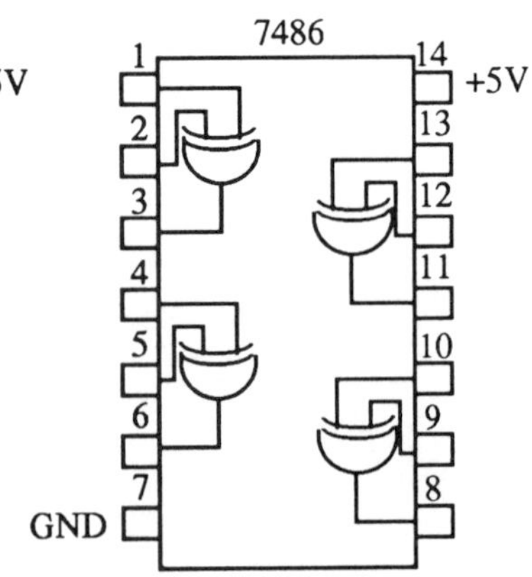

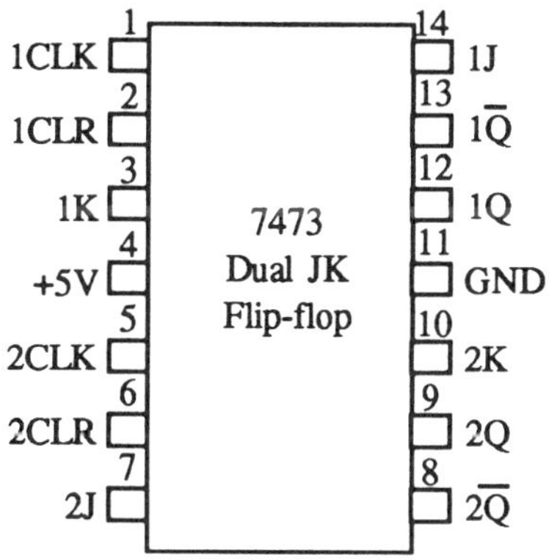
7473
Dual JK
Flip-flop
1 1CLK
2 1CLR
3 1K
4 +5V
5 2CLK
6 2CLR
7 2J
14 1J
13 $1\overline{Q}$
12 1Q
11 GND
10 2K
9 2Q
8 $2\overline{Q}$

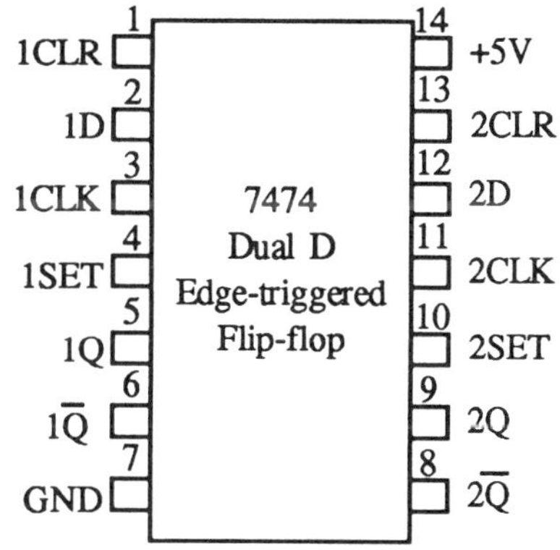
7474
Dual D
Edge-triggered
Flip-flop
1 1CLR
2 1D
3 1CLK
4 1SET
5 1Q
6 $1\overline{Q}$
7 GND
14 +5V
13 2CLR
12 2D
11 2CLK
10 2SET
9 2Q
8 $2\overline{Q}$

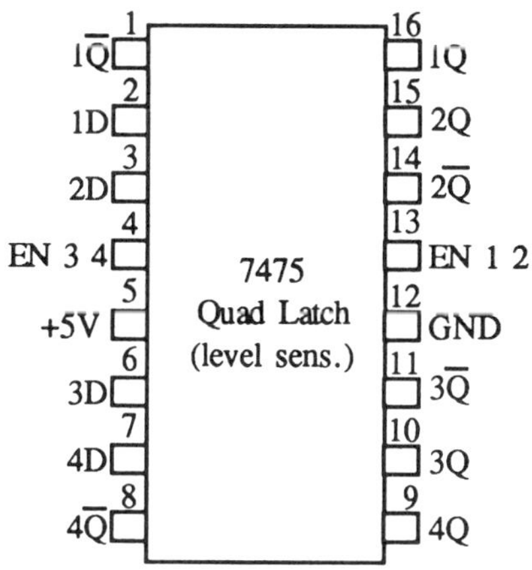
7475
Quad Latch
(level sens.)
1 $1\overline{Q}$
2 1D
3 2D
4 EN 3 4
5 +5V
6 3D
7 4D
8 $4\overline{Q}$
16 1Q
15 2Q
14 $2\overline{Q}$
13 EN 1 2
12 GND
11 $3\overline{Q}$
10 3Q
9 4Q

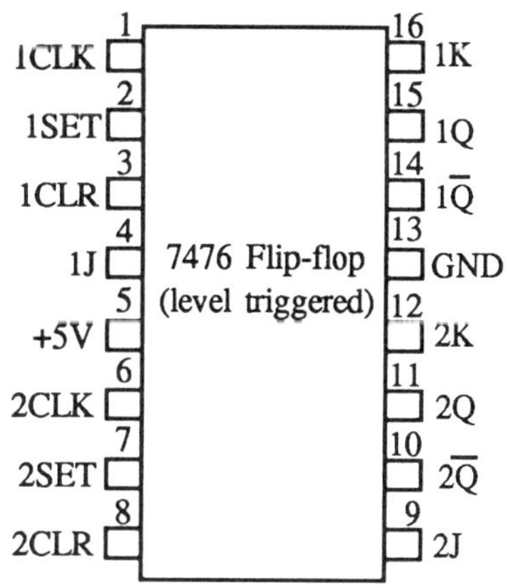
7476 Flip-flop
(level triggered)
1 1CLK
2 1SET
3 1CLR
4 1J
5 +5V
6 2CLK
7 2SET
8 2CLR
16 1K
15 1Q
14 $1\overline{Q}$
13 GND
12 2K
11 2Q
10 $2\overline{Q}$
9 2J

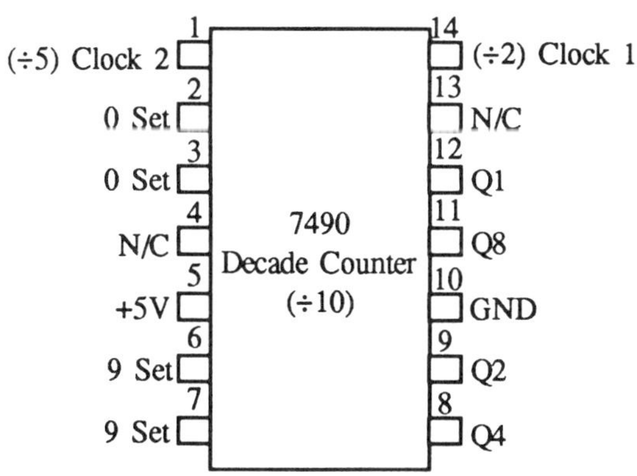
7490
Decade Counter
(÷10)
1 (÷5) Clock 2
2 0 Set
3 0 Set
4 N/C
5 +5V
6 9 Set
7 9 Set
14 (÷2) Clock 1
13 N/C
12 Q1
11 Q8
10 GND
9 Q2
8 Q4

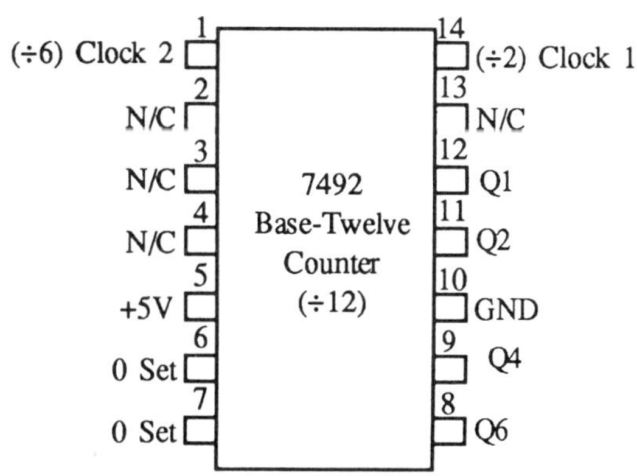
7492
Base-Twelve
Counter
(÷12)
1 (÷6) Clock 2
2 N/C
3 N/C
4 N/C
5 +5V
6 0 Set
7 0 Set
14 (÷2) Clock 1
13 N/C
12 Q1
11 Q2
10 GND
9 Q4
8 Q6

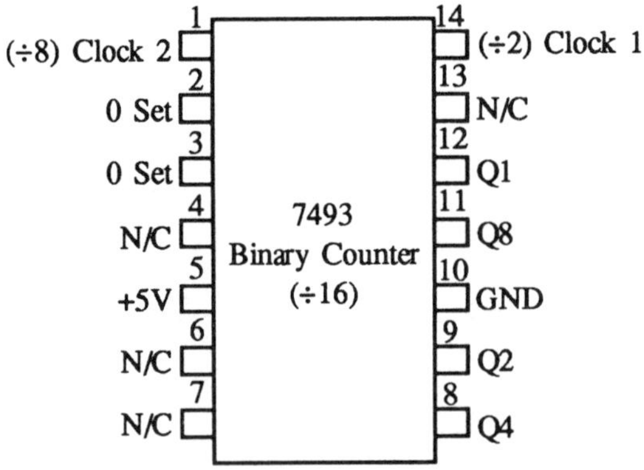

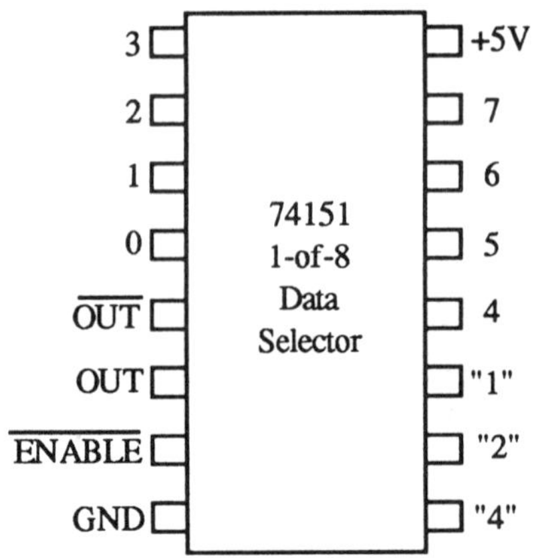

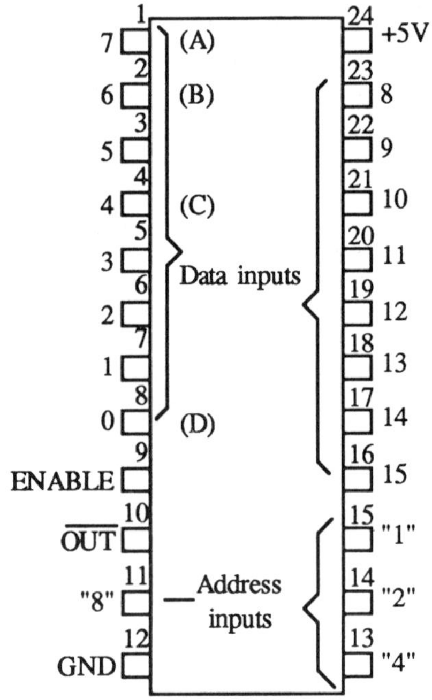

74150
1-of-16 Data
Selector

74154
1-of-16 Data
Distributor

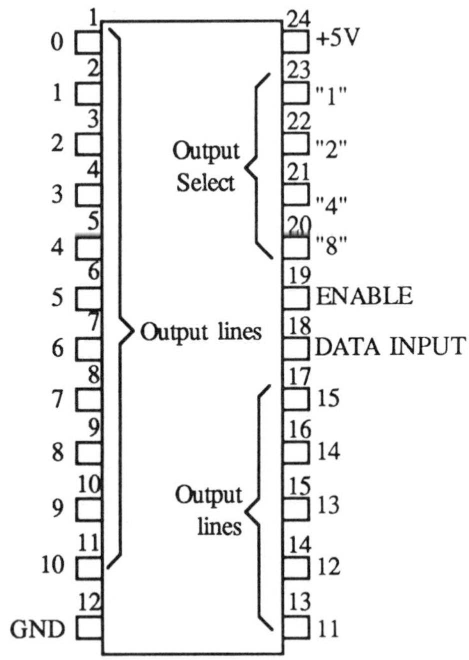

CMOS Integrated Circuits

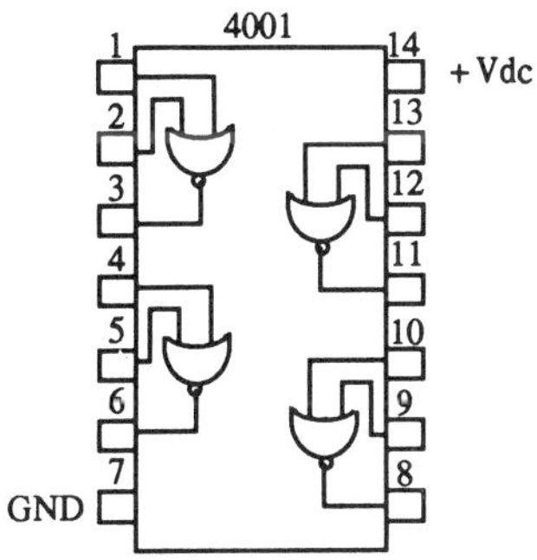

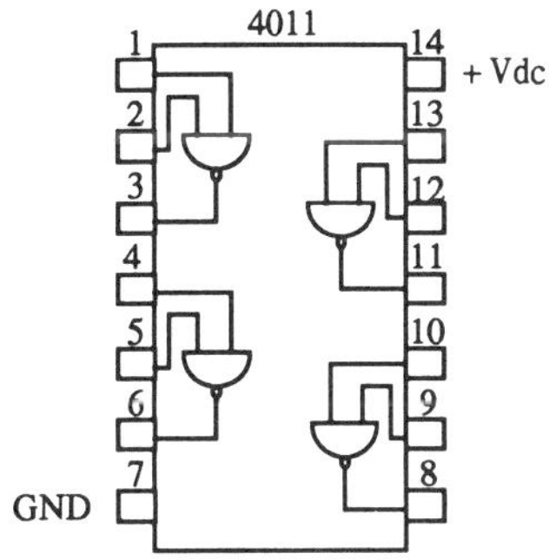

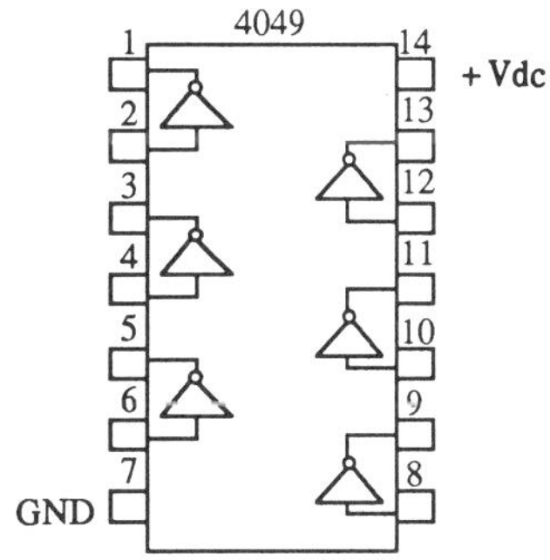

(Same as 4009)

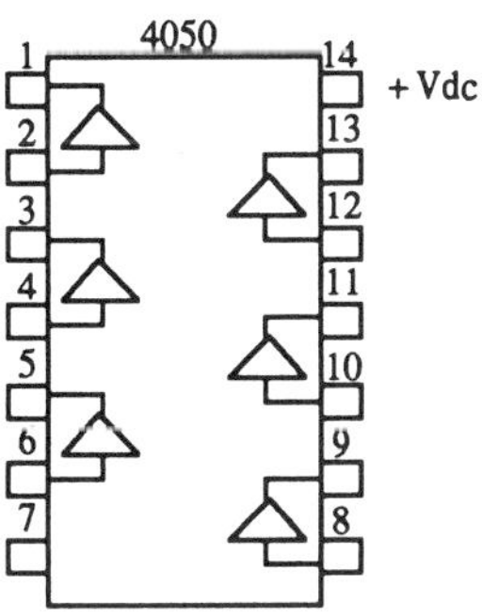

(Same as 4010)

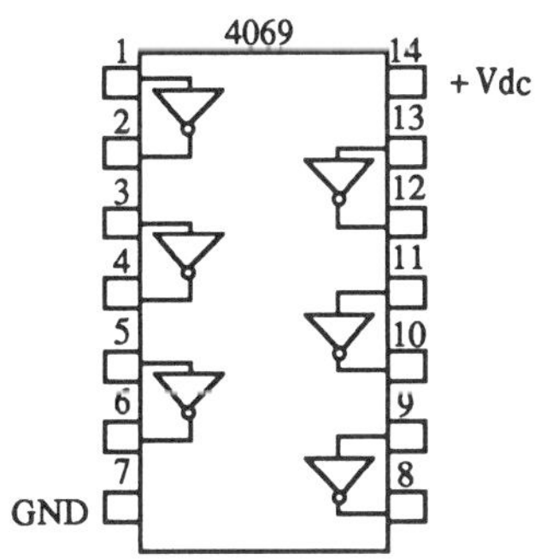

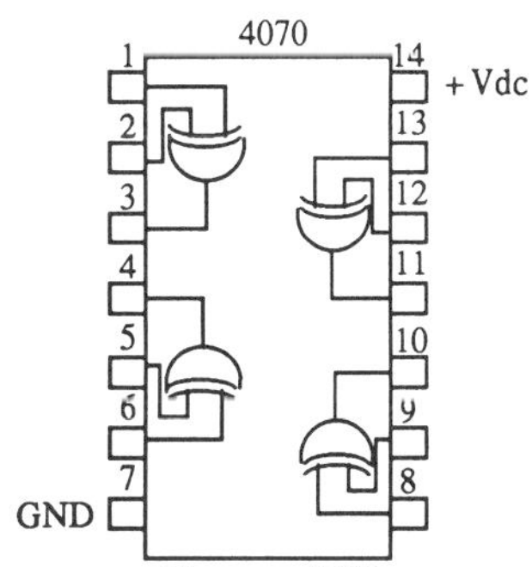

(Same as 4030)

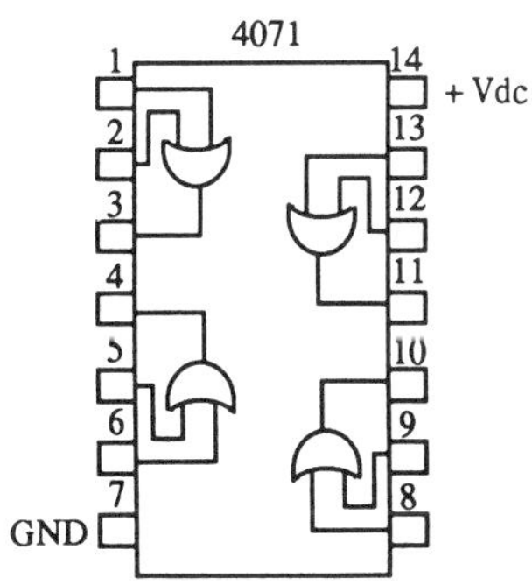

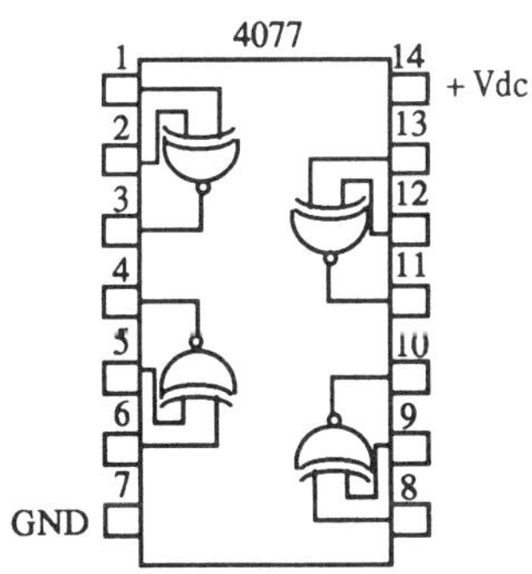

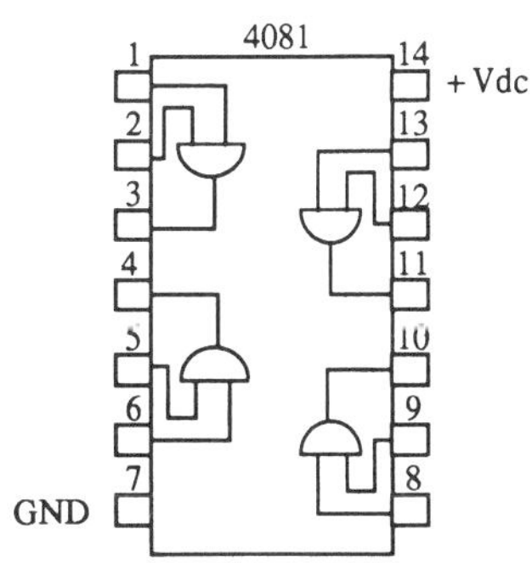

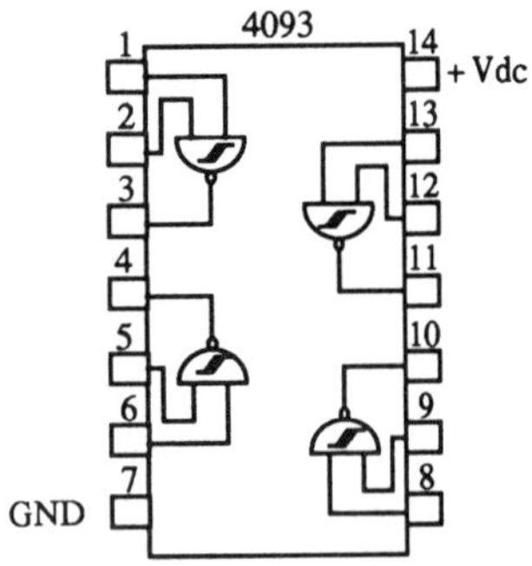

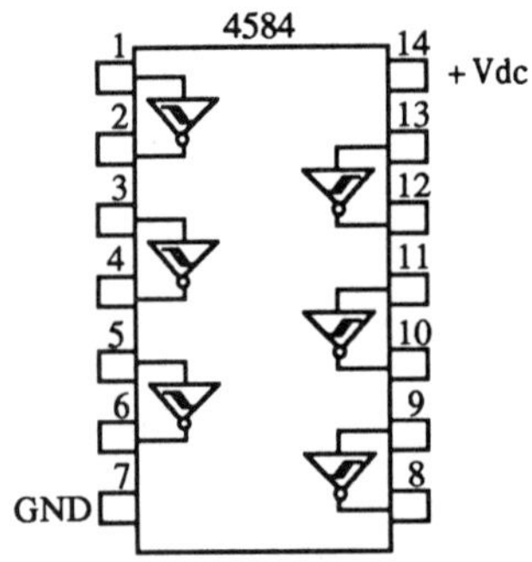

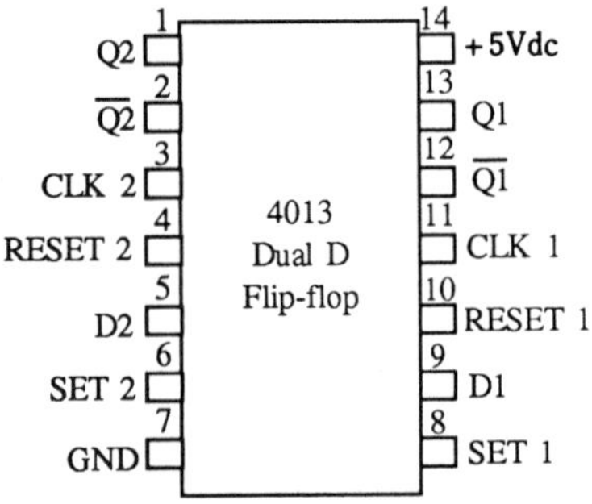

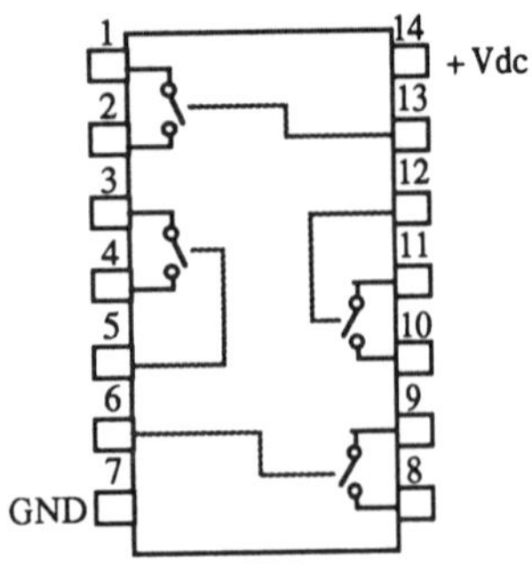

4016
Quad Digital
or Analog
Switch
(Same as 4066)

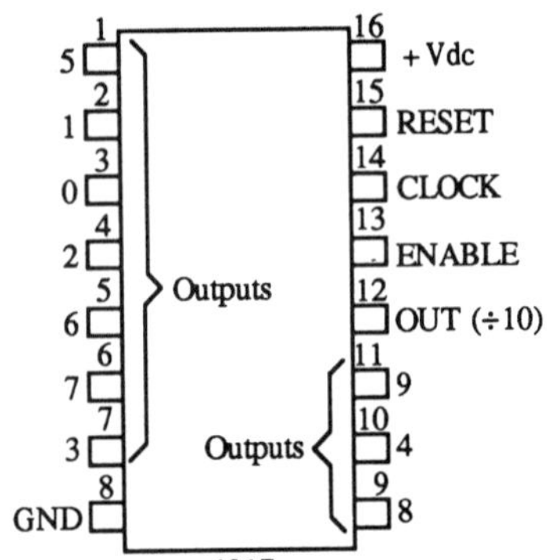

4017
Divide-by-10
Decade
Counter

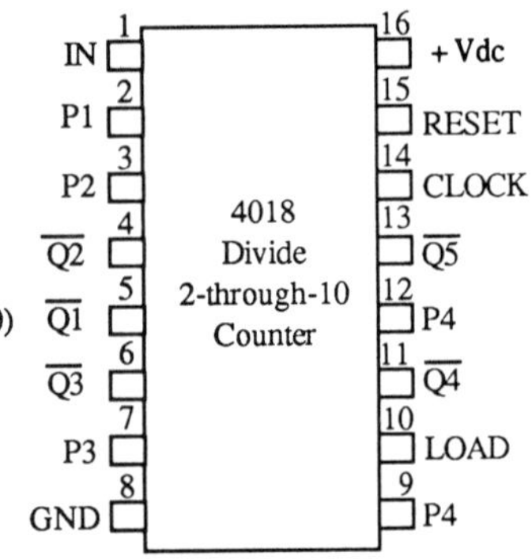

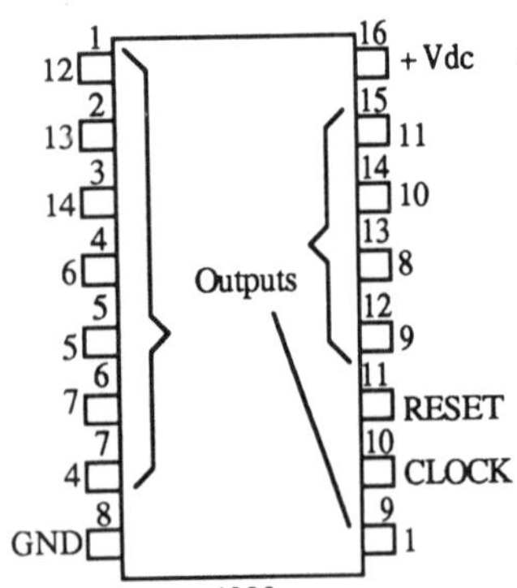

4020
14-state
(÷16,384)
Binary Counter

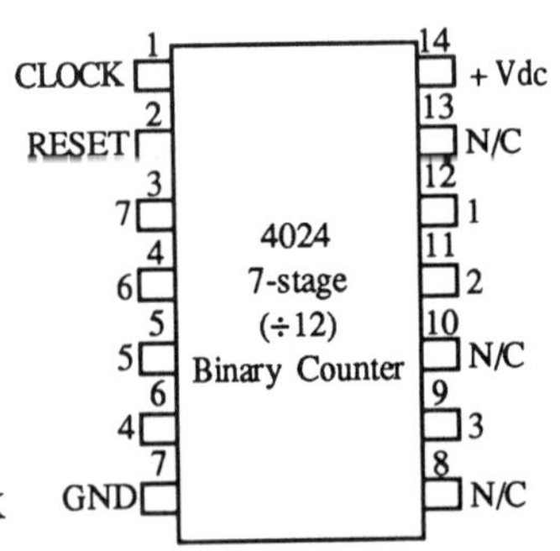

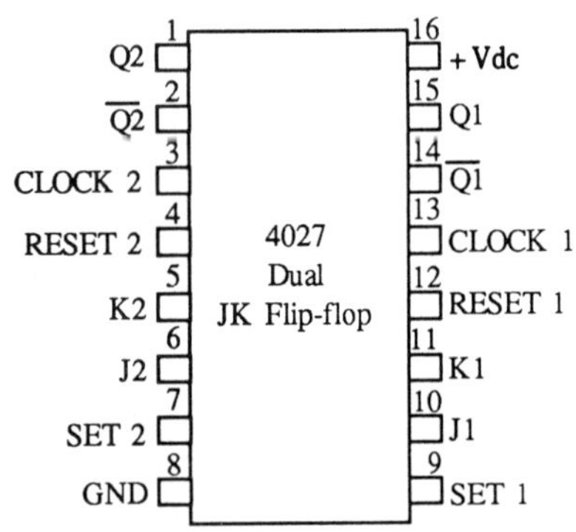

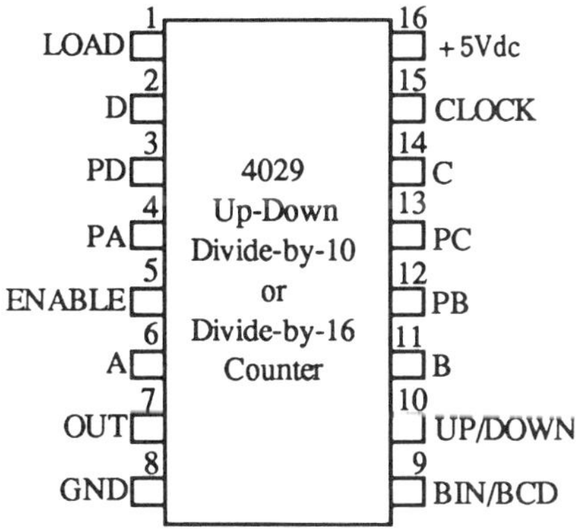

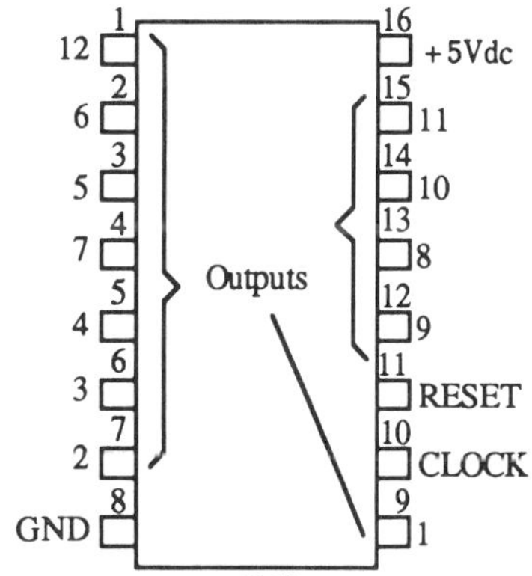

4040
12-Stage
(÷4,096)
Binary Counter

Linear Integrated Circuits

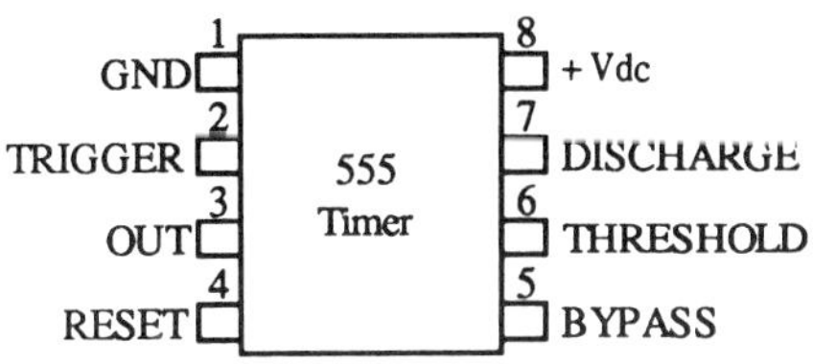

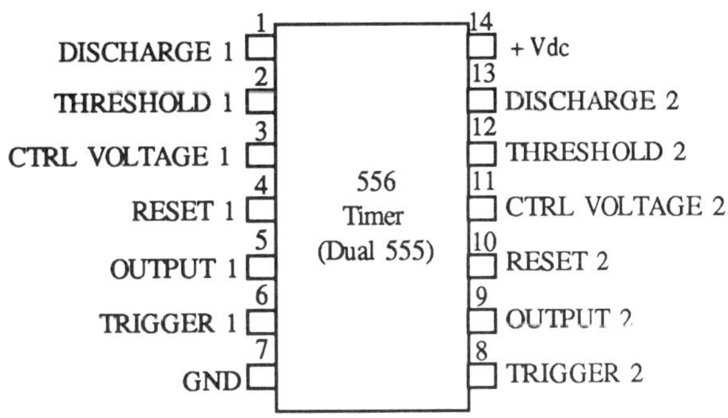

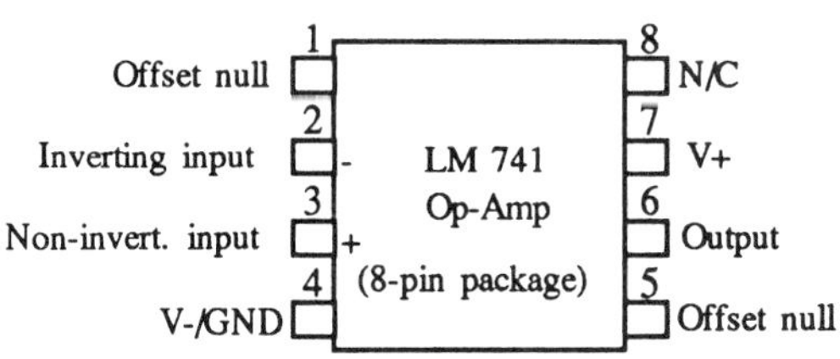

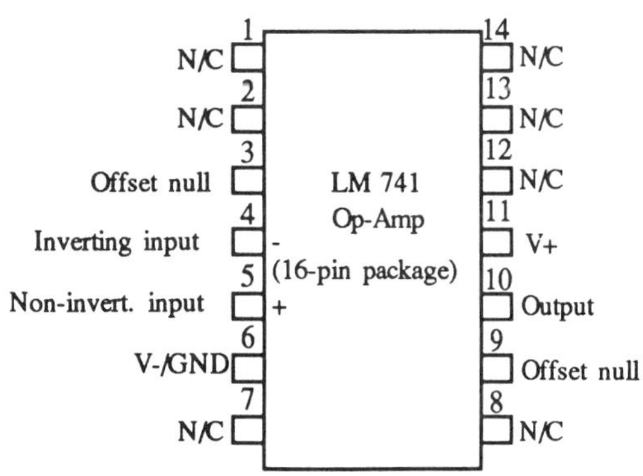

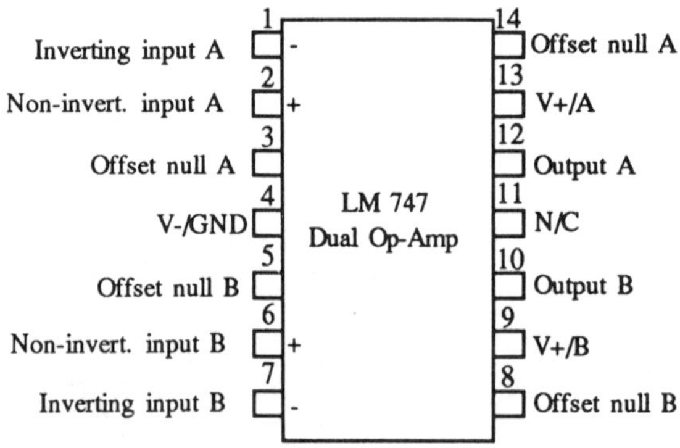
LM 747
Dual Op-Amp
1 Inverting input A
2 Non-invert. input A
3 Offset null A
4 V-/GND
5 Offset null B
6 Non-invert. input B
7 Inverting input B
8 Offset null B
9 V+/B
10 Output B
11 N/C
12 Output A
13 V+/A
14 Offset null A

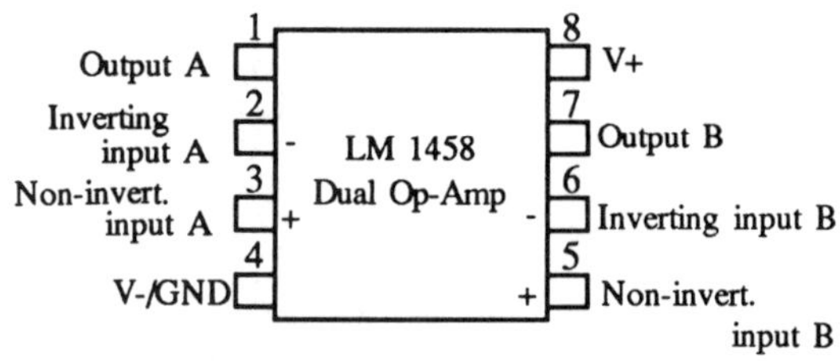
LM 1458
Dual Op-Amp
1 Output A
2 Inverting input A
3 Non-invert. input A
4 V-/GND
5 Non-invert. input B
6 Inverting input B
7 Output B
8 V+

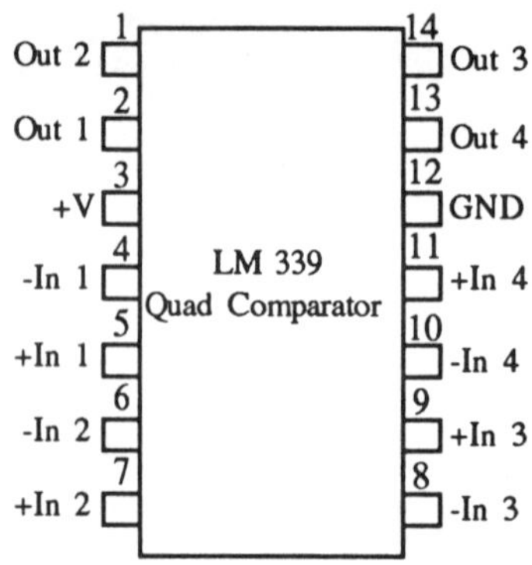
LM 339
Quad Comparator
1 Out 2
2 Out 1
3 +V
4 -In 1
5 +In 1
6 -In 2
7 +In 2
8 -In 3
9 +In 3
10 -In 4
11 +In 4
12 GND
13 Out 4
14 Out 3

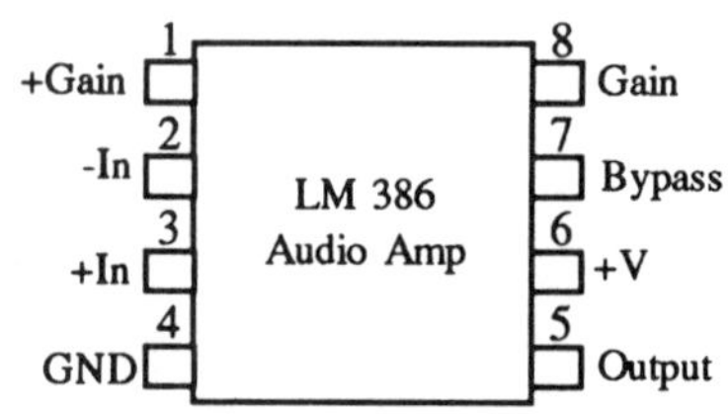
LM 386
Audio Amp
1 +Gain
2 -In
3 +In
4 GND
5 Output
6 +V
7 Bypass
8 Gain

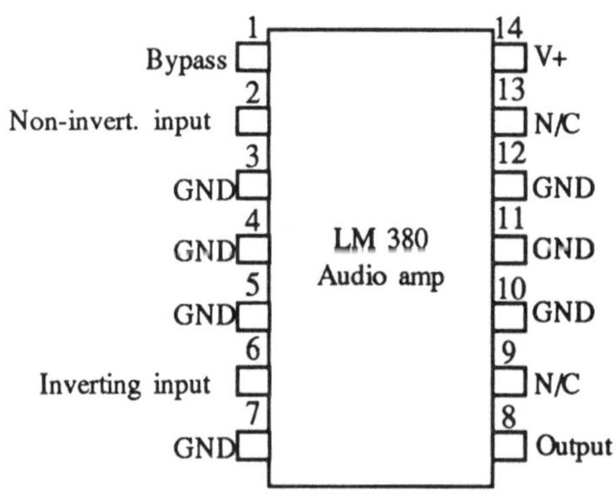
LM 380
Audio amp
1 Bypass
2 Non-invert. input
3 GND
4 GND
5 GND
6 Inverting input
7 GND
8 Output
9 N/C
10 GND
11 GND
12 GND
13 N/C
14 V+

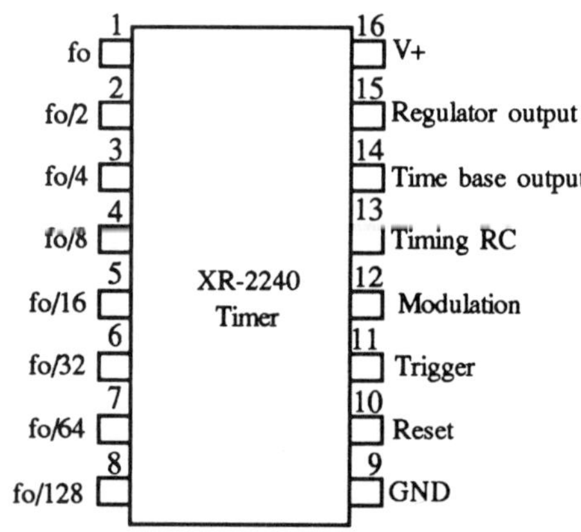
XR-2240
Timer
1 fo
2 fo/2
3 fo/4
4 fo/8
5 fo/16
6 fo/32
7 fo/64
8 fo/128
9 GND
10 Reset
11 Trigger
12 Modulation
13 Timing RC
14 Time base output
15 Regulator output
16 V+

11

Soldering tips and techniques

FEW ELECTRONIC PROJECTS CAN BE ASSEMBLED WITHOUT SOLDERING WIRES together. Soldering sounds and looks simple enough, but there really is a science about it. If you are unfamiliar with soldering or want a quick refresher course, read this primer on soldering fundamentals.

Tools and Equipment

Good soldering requires the proper tools. If you don't have them already, they can be purchased at Radio Shack or almost any electronics store. Let's take a quick look at the tools you will need.

Soldering iron

You'll need a soldering iron, of course, but not just any old soldering iron. Get a soldering "pencil" with a low-wattage heating element. For electronics work, the heating element should not be higher than about 30 watts. Most soldering pencils are designed so that you can change heating elements as easy as changing a light bulb. Avoid using a soldering iron that is not grounded, or you risk damaging sensitive electronic components by subjecting them to electrostatic discharge (see later in this chapter).

Do not use the instant-on type of soldering guns, favored in the old tube days. They create far too much unregulated heat, and they are too large to effectively solder most joints on a printed circuit board.

If your soldering iron has a temperature control and readout, dial it to between 665 and 680 degrees. This provides maximum heat with the minimum danger of damage to the electronic components. If your iron has just the heat control but lacks the readout, initially set it to low. Wait a few minutes for the iron to heat up, then try one or two test connections. Adjust the heat control so that solder flows onto the connection in under five seconds. Longer times mean that the control might be set too low, and the prolonged heat could damage components and circuit boards. Conversely, if the solder melts instantaneously and gives off copious amounts of fumes, the control might be set too high. Reduce the setting, wait a few seconds for the iron to cool, and try another connection.

When you are not using your soldering iron, keep it in an insulated stand, as shown in Fig. 11-1. Don't rest the iron in an ashtray or precariously on the carpet. You or some precious belonging is sure to be burned.

Fig. 11-1. A soldering iron in its stand.

Soldering tips

The choice of soldering tip is important. For best results, use a fine tip designed specifically for printed circuit board use (unless you are soldering larger wires, in which case use a larger tip). Figure 11-2 shows some common tips for use with a soldering pencil. Tips are made to fit certain types and brands of heating elements, so make sure you get the kind for your iron.

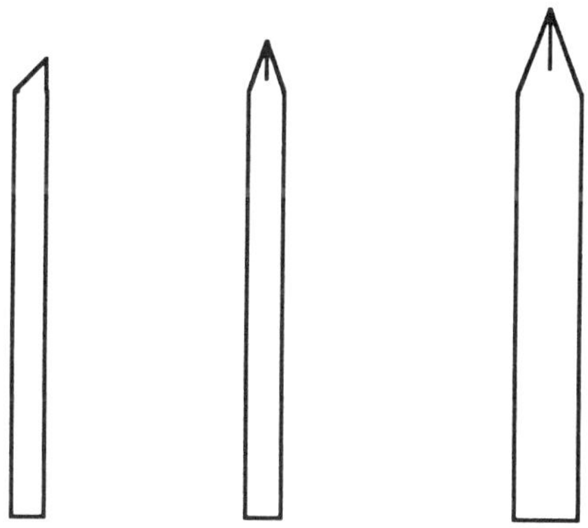

Fig. 11-2. A sampler of soldering tips. Use the right one for the job.

If the tip doesn't come pretinned, tin it by attaching the tip to the iron, and heat the iron to its normal temperature. When the iron is hot, apply a thin coat of solder to the entire tip. With a damp sponge (see below), wipe off the excess solder so that only a thin, silvery coating remains on the tip. Periodically retin the tip of your soldering iron to keep it bright and shiny.

Sponge

Keep a damp sponge (just about any type of kitchen sponge will do) by the soldering station and use it to wipe off extra solder. Do not allow globs of solder to remain on the tip; the glob could come off while you're soldering and ruin the connection. The excess solder can also draw heat away from the tip of the pencil, causing a poor soldering job. You'll have to rewet the sponge now and then if you are doing lots of soldering. Never wipe off solder onto a dry sponge because the sponge could melt or catch on fire. Used daily, the sponge should last several weeks. Replace it when it becomes worn out.

Solder

You should only use rosin-core solder. It comes in different thicknesses; for best results, use the thin type (0.050 inch) for most electronics work, especially if you're working with printed circuit boards. *Never* use acid core or silver solder on electronic equipment.

Solder should be kept clean and dry. Avoid tossing your spool of solder into your electronics junk bin. It can collect dust, grime, oil, grease, and other contaminants. Dirty solder requires more heat to melt. In addition, the grime fuses with the solder and melds into the connection, which can jeapordize the integrity of the joint. If your solder becomes dirty, wipe it off with a damp paper towel soaked in alcohol and let dry.

Soldering tools

Basic soldering tools include a good pair of small needle-nosed pliers, tweezers, wire strippers and wire cutters (sometimes called ''side'' or ''diagonal'' cutters). The stripper should have a dial that lets you select the gauge of wire you are using. A pair of ''nippy'' cutters, which cut wire leads flush to the surface of the board, are handy but not absolutely essential.

A heatsink is used to draw heat away from components during soldering. To use the heatsink, clip it onto the lead of the component you are soldering. For best results, the heatsink should be placed as close to the component as possible. Even with a heatsink, however, you must still exercise caution and avoid overheating components.

Cleaning supplies

Cleaning is often necessary before and after soldering. Isopropyl alcohol makes a good, all-around soldering cleaner. You can also use contact cleaner (without lubrication). After soldering, and when the components and board are cool, spray or brush on some flux remover.

Solder vacuum

A solder vacuum is a suction device that is used to pick up excess solder. It is often used when desoldering (removing a wire or component on the board). Solder can also be removed using a length of copper braid; most electronics stores sell braid in spools.

Board and Component Lead Cleaning

Before soldering, make sure all parts of the connection are clean. If you're soldering a component onto a printed circuit board (especially if you've just made the board using one of the techniques outlined in chapter 14), clean the board first by following these steps:

- Fill a bowl with lukewarm water.
- Add just a drop or two of liquid dish detergent. Mix the detergent to produce a small amount of suds.
- Dip the board into the bowl and scrub the copper (or tin-plated) traces with a nonmetallic scrubbing pad (like 3M Scotchbrite). Do not rub too vigorously.
- Rinse under cold water and blot dry with a paper towel. If desired, use compressed air to drive the trapped water out of the component holes. Let the board dry completely for 10 to 15 minutes.

When you are ready for soldering, wet a cotton ball with isopropyl alcohol (or other cleaner) and wipe off all connection points. Wait a few seconds for the alcohol to evaporate, then commence soldering.

Soldering Basics

The basis of successful soldering is that the soldering iron is used to heat up the work, whether it's a component lead, a wire, or whatever. You then apply the solder to the work, as shown in Fig. 11-3. *Do not* apply solder directly to the soldering iron. If you take the shortcut by melting the solder on the iron, you might end up with a ''cold'' solder joint. A cold joint doesn't adhere well to the metal surfaces of the part or board, so electrical connection is often impaired.

Once the solder flows around the joint (and some will flow to the soldering iron tip), remove the iron and let the joint cool. Avoid disturbing the solder as it cools; a cold joint might result. Do not apply heat any longer than necessary. Prolonged heat can permanently ruin electronic components. A good rule of thumb is that if the iron is on any one spot for more than five seconds, it's too long.

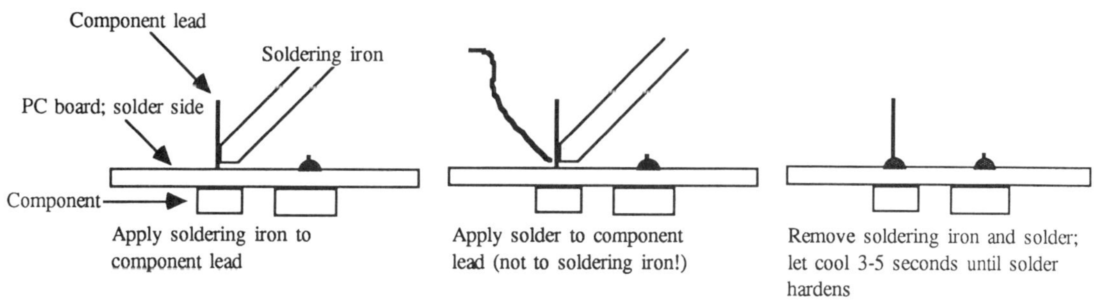

Fig. 11-3. Always solder in the proper sequence.

If at all possible, keep the iron at a 30- to 40- degree angle to work, as shown in Fig. 11-4. Most tips have a beveled tip for this purpose.

Apply only as much solder to the joint as is required to coat the lead and circuit board pad. A heavy-handed soldering job can lead to soldering bridges (see Fig. 11-5), which is when one joint melds with joints around it. At best, solder bridges prevent the circuit from working; at worse, they cause short circuits that can burn out the entire board.

When the joint is complete and has cooled, test it to make sure that it is secure. Wiggle the component to see if the joint is solid. If the lead moves within the joint, you'll have to resolder it (the joint might be dirty, so you might have to remove the old solder first, clean the connection, and try again).

When soldering on printed circuit boards, you'll need to clip off the excess

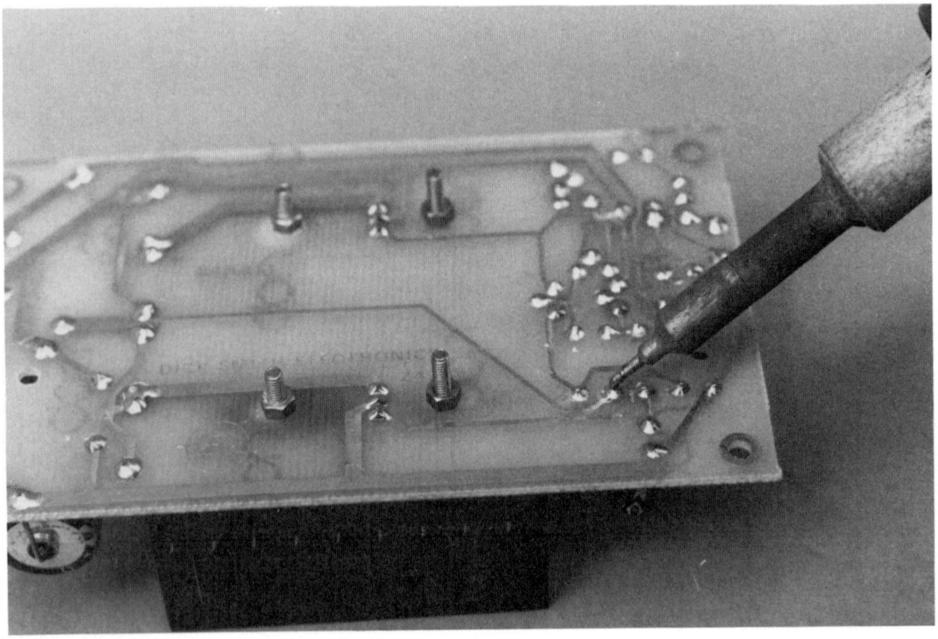

Fig. 11-4. Hold the soldering iron or pencil at a 30- to 40-degree angle to the work, and never apply solder directly to the tip of the iron.

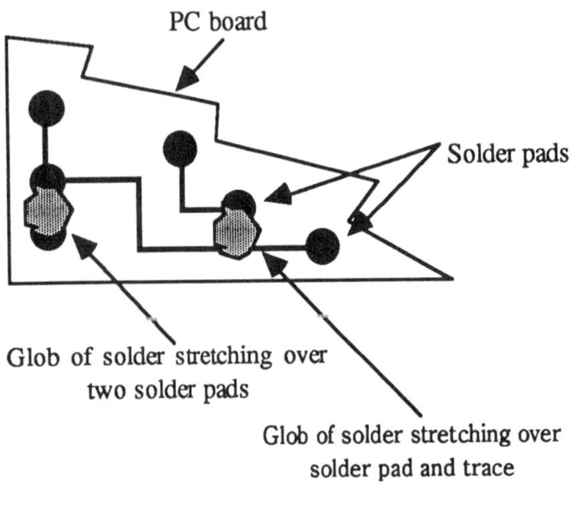

Fig. 11-5. A solder bridge.

leads that protrude beyond the solder joint. Use a pair of diagonal or nippy cutters for this task, as shown in Fig. 11-6. Be sure to protect your eyes when cutting the lead; a bit of metal could fly off and lodge in one of your eyes.

You might want to wear protective eyewear, like the plastic nonprescription glasses illustrated in Fig. 11-7. These glasses protect your eyes from flying debris, yet they afford good visibility for exacting work. If you wear glasses, you'll need goggles that fit over them, or else use prescription safety glasses.

Fig. 11-6. Nipping the excess lead after soldering.

Fig. 11-7. Safety glasses.

A Good Solder Joint

A good solder joint should be bright and shiny. A joint that looks dull is probably cold and should be remade. The joint should not have any sharp ''peaks.'' If so, the solder didn't flow well enough to make a good connection. Remake the joint, being sure to apply the iron to the work and not to the iron. Excess solder that forms on the tip (another cause for the ''peaks'') should be removed using the damp sponge.

Get into the habit of inspecting all of your solder joints. Use a magnifying glass to highlight minute detail. Be on the lookout for:

- Air bubbles
- Incomplete joints
- Solder bridges
- Peaks in solder joints
- Weak solder joints (the solder flakes off easily)

Replacing Components

Replacing a soldered component first requires that you remove all the solder holding it in place. Use the soldering iron to melt the joint, and as the solder flows, suck it up with a solder vacuum or wick it up with a copper braid (see Fig. 11-8). When using copper braid, cut off the portion that has absorbed the excess solder and avoid leaving the braid in place when you remove the iron from the connection. Avoid touching the braid—it's hot!

Fig. 11-8. Copper braid can be used to remove excess solder or when desoldering components from a circuit board.

You will need to remove enough solder so that the component lead is free. If you can't get all the solder up the first time, let the joint cool (this is very important) and try again.

Clean the old joint and the leads of the replacement component in alcohol to remove any oil that might impair the grip of the joint after soldering.

Gently insert the new component. Don't pull on the lead or you could damage the component. Once the part is seated on the board, bend the leads slightly to keep it in place. Solder as usual. If you are resoldering a wire onto a terminal, wrap the stripped end of the wire around the eyelet of the terminal prior to soldering.

Electrostatic Discharge

Electrostatic discharge—better known as a ''carpet shock''—can ruin electronic components. Remove the excess static buildup from your body by touching some grounded metal object prior to soldering or before handling electronic parts and boards. If you are soldering transistors and integrated circuits, use a grounded soldering iron as well as an antistatic wrist band and table mat. Refer to chapter 6, ''Eliminating Static Electricity,'' for more details on combating electrostatic discharge.

Iron Tip Maintenance and Cleanup

After soldering, let the iron cool. Loosen the tip from the heating element (use a pair of pliers and grip the tip by the shaft) and store it for next use. After several soldering sessions, the tip should be cleaned using a soft brush. Don't file it or sand it down with emory paper.

After many hours of use, the soldering tip will become old, pitted, and deformed, at which time it should be replaced. Old or damaged tips impair the transfer of heat from the iron to the connection, and that can lead to poor soldering joints. Be sure to replace the tip with one made for your soldering iron. Tips are generally not interchangeable between brands.

Invariably, little nuggets of solder remain after a solder job. Make sure that these balls of once-molten solder are not left on the printed circuit board or inside your project. The solder could bridge wires or board traces together and cause a serious short circuit. Inspect your work carefully and use a soft brush to whisk away stray bits of solder.

Solder Safety

Although these safety precautions were mentioned in chapter 5, they are worth repeating here.

- A hot soldering iron can seriously burn you. Keep fingers away from the tip of the soldering pencil.
- Never touch a solder joint until it has cooled.
- During use, always place the soldering iron in a properly designed iron stand.

- While fumes produced during soldering are not particularly offensive, they are mildly toxic. Avoid inhaling the fumes for any length of time. Work only in a well-ventilated area.
- Protect your eyes when clipping leads off components. If possible, wear eye protection like safety glasses or optically clear goggles.
- Keep a fire extinguisher handy, just in case. Halon extinguishers ''snuff'' out the flames rather than cover up circuits with quenching powder like other chemical extinguishers, but remember that Halon displaces oxygen and poses a serious suffocation hazard.

12

Solderless and solder breadboards

SOLDERLESS BREADBOARDS (ALSO CALLED EXPERIMENTER'S BREADBOARDS) enable you to experiment with circuits without having to solder components together. Unless you are building a project from a ready-to-go kit, you will want to test each of the circuits you build on a solderless breadboard. In this way, you can more easily correct for mistakes, as well as readily modify the circuit to suit your requirements.

Let's take a closer look at using solderless breadboards in this chapter. You'll also learn how to convert your experimental circuits from temporary breadboard to permanent printed circuit.

Anatomy of a Solderless Breadboard

Solderless breadboards are not designed for permanent circuits. Rather, they are engineered so that you can try out a circuit and experiment with it, without the trouble of wiring and soldering. Then, when you are assured that the circuit works, you can use one of the many construction techniques available.

You are urged to first test all the circuits you want to build on a solderless breadboard. You'll find that you can often improve on the performance of the circuit just by changing a few component values. Such changes are easiest when you can simply remove one component—without unsoldering and resoldering—and exchange it for another.

A solderless breadboard, like the basic model shown in Fig. 12-1, consists of a series of holes with internal contacts. These contacts are spaced one-tenth

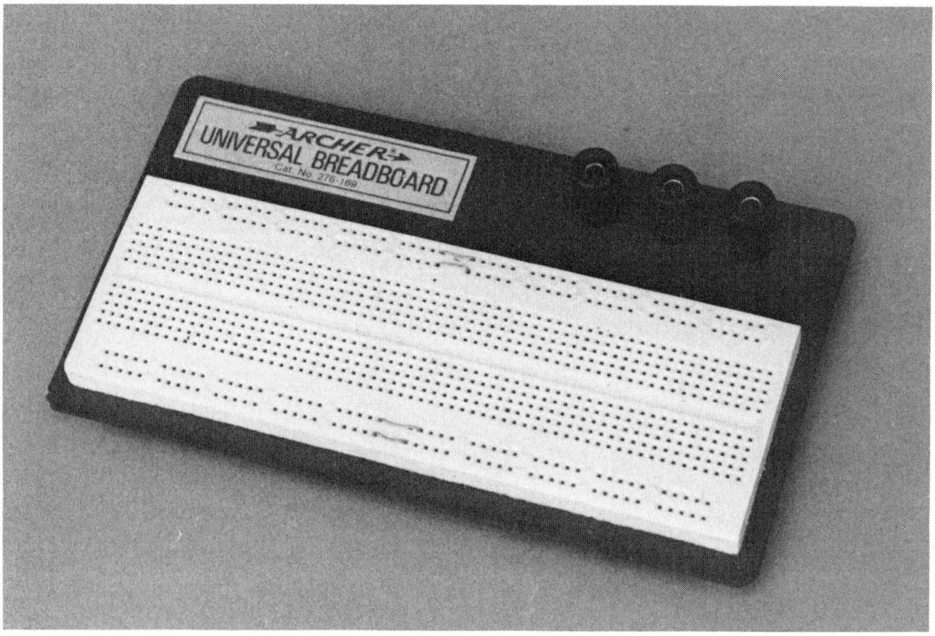

Fig. 12-1. Sample single-width solderless breadboard.

of an inch apart—just right for integrated circuits, most transistors, and discrete components like capacitors and resistors. You merely plug in ICs, resistors, capacitors, transistors, and 20- or 22-gauge wire in the proper contact holes to create your circuit.

The contacts are wired together to form multiple-point contacts, as shown in Fig. 12-2. The basic solderless breadboard contains 550 contact points, organized into two rows of 47 columns each. Two additional, longer rows are used as common tie points for positive and negative power supply. You can connect as many as five wires or components together along a single column.

The contact points are usually made from a springy beryllium-copper metal coated with silver-nickel. The plating prevents the contacts from oxidizing and the springiness of the metal allows you to use different diameter wires and component leads without seriously deforming the contacts. However, the contacts can be damaged if you attempt to use wire that's larger than 20 gauge.

Solderless breadboards come in many sizes. Breadboards with 550 contact points can accommodate designs with about three or four 14- or 16-pin ICs. For the most flexibility, get a "double-width" board, as illustrated in Fig. 12-3, one that can accommodate at least 10 ICs. These contain over 1,200 contact points. For elaborate design work, you can purchase extra-large breadboards that contain 3,200 contacts or more.

While you're buying a breadboard, purchase a set of prestripped wires. The wires come in a variety of lengths, and are already stripped and bent for use in

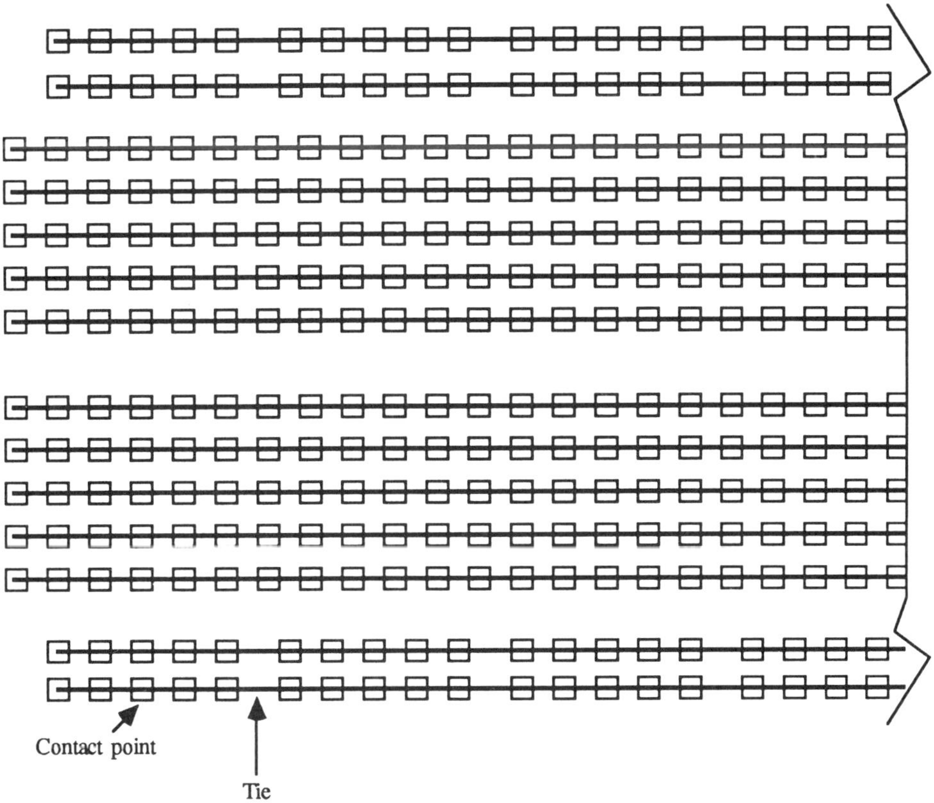

Fig. 12-2. Contact layout single-width breadboard.

breadboards. The set costs $5 to $7, but you can bet the price is well worth the time you'll save. If you decide to make your own set of breadboard wires, use solid 22-gauge insulated wire. Cut the wires into the following lengths:

Length	Quantity
3/4 inch	10
1 inch	20
1 1/4 inch	25
1 1/2 inch	25
2 inch	10
2 1/2 inch	10
3 inch	10
4 inch	5
6 inch	5

Once cut, strip 1/4 inch of insulation from both ends of the wire, as shown in Fig. 12-4. While stripping the insulation, insert one end of the wire into the

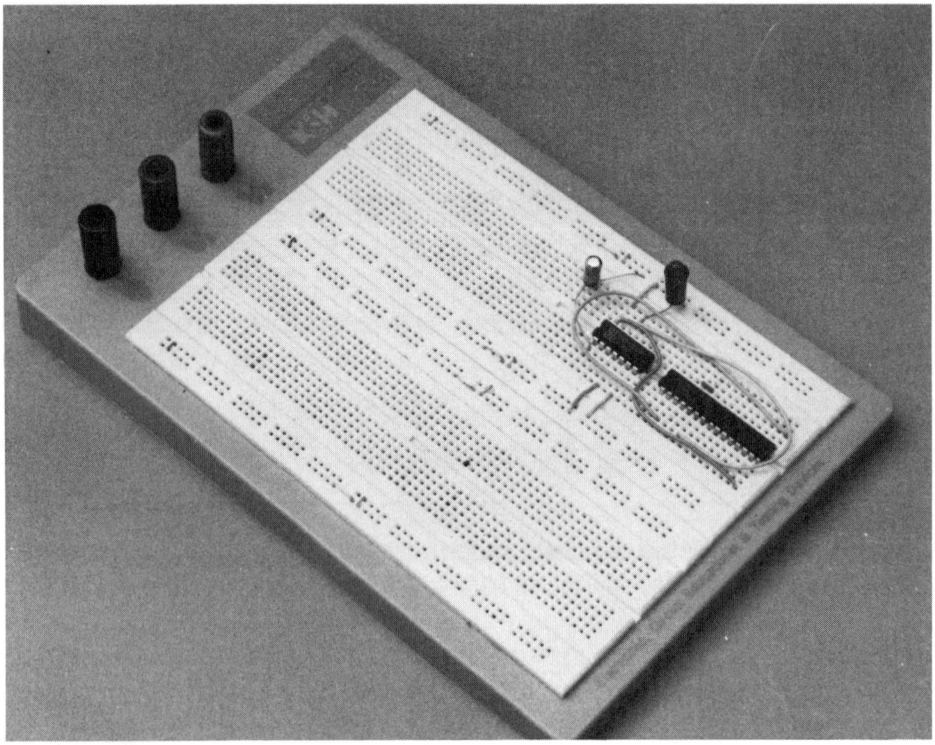

Fig. 12-3. Sample double-width solderless breadboard.

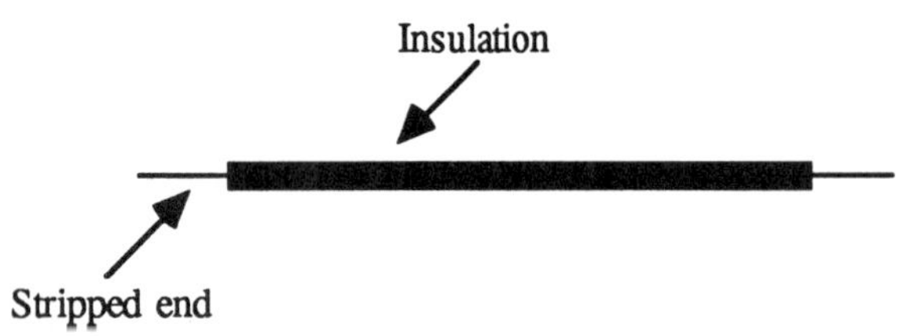

Fig. 12-4. Sample stripped wire.

stripping tool, and hold the other end with a pair of needle-nosed pliers. If you have an automatic wire stripper/cutter, you can cut the wire and strip the insulation in one easy step. After stripping the insulation, use a pair of needle-nosed pliers to bend the exposed ends of the wire at 90 degrees, as shown in Fig. 12-5.

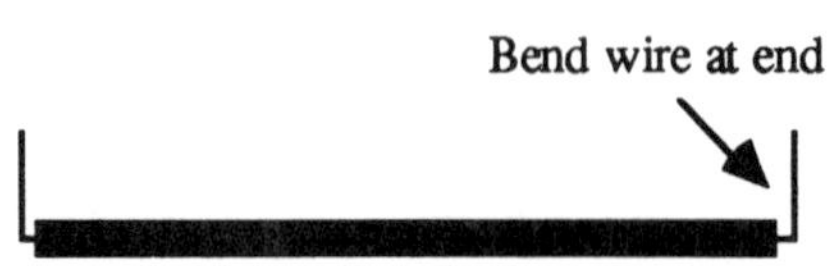

Fig. 12-5. Wire bent at ends.

Here are some tips to keep in mind when using a solderboard:

- Use a chip inserter/extractor to implant and remove ICs. This reduces the chance of damaging the IC during handling. If you are working with CMOS chips, ground the inserter/extractor tool to eliminate stray static electricity.
- When using CMOS chips, build the rest of the circuit first, using a "dummy" TTL IC to make sure you wire everything properly. Be sure to provide connections for the positive and negative power supply, and verify that all inputs are connected to something (those that are not used should be tied to the positive or negative supply rail). When you are ready to test the circuit, remove the dummy chip and replace it with the CMOS IC.
- When inserting wire, use small, needle-nosed pliers to plug the end of the wire into the contact hole. If the wire is short, use the pliers to gently pull it out of the hole when you are done with the breadboard.
- Never expose a breadboard to heat, or the plastic could be permanently damaged. ICs and other components that become very hot (because of a short circuit or excess current, for example) can melt the plastic underneath them. Check the temperature of all components while the circuit is under power.
- Solderless breadboards are designed for low-voltage dc experiments. They are not designed, nor are they safe, to carry 117 Vac house current.
- Avoid building a bird's nest where the connection wires are routed carelessly. This makes it harder to debug the circuit and it greatly increases the chance of mistakes. Worse, the tangle of wires can cause the circuit to malfunction altogether. Your breadboard circuits should be carefully planned and constructed. This requires more time and patience on your part, but after building a few projects, you'll find the extra effort is well worth it.

Permanent Circuits on Solder Breadboards

Akin to the solderless breadboard is the *solder breadboard*, also called a solderboard, an experimenter's PC board, or a universal solderboard. The solder breadboard allows you to make permanent any design you create on a solderless breadboard.

The solderboard comes preeteched with the same tie points as the 550-contact solderless breadboard. Simply solder the components into place, using jumper wires to connect components that can't be directly tied together.

If the circuit only takes up a portion of the entire length of the solderboard, cut off the extra using a hacksaw or razorsaw. Because you've already built the circuit on a solderless breadboard, you know exactly how much space is required on the solderboard. You can cut the solderboard to size before you even mount any components or wires onto it, but don't forget to leave space in the corners of the board to drill new mounting holes so you can secure the board inside whatever enclosure you are using. Alternatively, you can secure the board to a frame or inside an enclosure using double-sided foam tape. The tape acts to cushion the board and prevents breakage, and the thickness of the foam prevents the underside of the board from actually touching the chassis.

The main disadvantage of universal solderboards is that they don't provide extremely efficient use of space. Unless you cram the components onto the board, you are limited to building circuits with only two or four ICs and a handful of discrete components.

Solderboards are available at most electronics stores, including Radio Shack. The cost is reasonable—usually under three dollars per board. If you're constructing small circuits, one solderboard will last several projects. In time, you'll learn how to conserve space to make good use of the real estate on a solderboard.

Occasionally, electronics magazines such as *Radio Electronics* or *Modern Electronics* publish a construction article using the 550-contact solderboard as a template. Such a circuit has an outline of a solderboard along with an overlay indicating where the components and wiring are to be placed. This method provides a fast and easy way of developing a simple printed circuit board for a project and eliminates the need to etch and drill a custom-made PCB.

For safety, you might want to build the circuit on a solderless breadboard first, to make sure everything works as it should. When you are satisfied that the circuit works as detailed in the article, you can transfer the component and wiring to the solderboard.

IMPORTANT NOTE: While most circuits you test with a solderless breadboard will work satisfactorily, some projects might behave erratically or unpredictably unless they are built using a printed circuit board and solder. This is particularly true of RF circuits, like miniature radio receivers and transmitters, and sensitive timing circuits that rely on exact component values. Solderless breadboards have a tendency to change the characteristics of some components—most notably capacitors and inductors—and this can change the way the circuit behaves.

Point-to-Point Perforated Board Construction

An alternative to the solderboard is point-to-point perforated circuit board construction. This technique refers to mounting the components on a predrilled board and connecting the leads together directly with solder. This technique was used extensively in the pre-IC days and was even found on commercial products.

With the proliferation of ICs, transistors, and other high-speed electronics, however, the point-to-point wiring method is all but abandoned.

Circuits that depend on close tolerances for timing and amplification cannot tolerate point-to-point wiring. Unless you are careful and use insulated wire, point-to-point construction invites short circuits and burnouts. You are advised not to use point-to-point perf board construction for any circuit that uses more than a dozen discrete components.

Using Predrilled Perf Grid Boards

A predrilled perf board with copper traces for wiring is a *grid board*, also called a universal, general-purpose, or prototyping PC board. Grid boards come in a variety of sizes and styles; a few are shown in Fig. 12-6. All are designed for use with ICs and other modern-day electronic components. One of the main applications of the grid board is circuit construction using wire wrapping (as detailed in chapter 15). But some designs can also be used to solder components directly onto the board.

Personally, I prefer the plain universal PC board with interleaved buses. Components are tied together using three point contacts (additional components can be linked together by bridging contacts). Between each contact are two strips for the negative and positive power supply. This layout is ideal for circuits

Fig. 12-6. Two popular grid boards. These are available at Radio Shack and elsewhere.

that use many integrated circuits. The interleaving of the power supply rails also helps reduce inductive and capacitive effects.

Use the grid board the same way as a solderboard. Plug the components into the board and solder them into place. Where possible, use the copper traces on the board to bridge components together—an IC pin connected to one leg of a resistor, for example. Then use insulated wire for connections that cannot be bridged locally.

13

Circuit board layout techniques

THE ELECTRONIC CONSTRUCTION TECHNIQUE OF CHOICE IS THE PRINTED circuit board, or PCB. There are a number of ways to build a custom printed circuit board for your project. This chapter details several methods including direct etch, photo transfer, heat/fusion transfer, and dry transfer. You'll also learn what makes up a circuit board and how to use a computer to speed up your work.

Anatomy of a Circuit Board

Before learning the various ways to make a printed circuit board, lets take a closer look at the design of the typical PCB.

Printed circuit boards consist of a thin copper sheet glued over a plastic, epoxy, or phenolic base, as shown in Fig. 13-1. This copper sheet is known as the *cladding* and represents the "foil" side of the board. Portions of the copper are then etched away, leaving just the printed circuit design. The exact method of laying out the circuit and etching the board is covered in more detail later in this chapter and in chapter 14, "Printed Circuit Board Etching."

The printed circuit consists of two primary parts: pads and traces. *Pads* are contact points for the components and are generally round ("donuts") or rectangular in shape. After the circuit board has been etched, a hole is drilled in the center of each pad. The components mount on the top of the board with their leads poking through the holes. Each component is then soldered to the

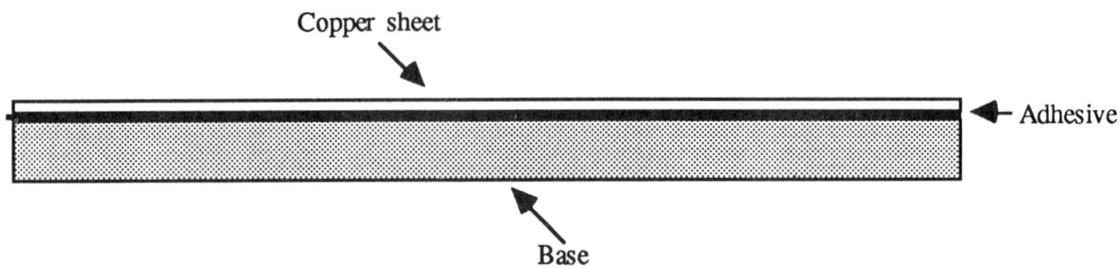

Fig. 13-1. Edgeview of a printed circuit board.

board at its pads. *Traces* (or ''tape'') are the wires of the circuit board. They stretch between the pads to electrically connect the components together.

Printed circuit boards can be either single- or double-sided. Single-sided boards have copper clad on only one side. The components are mounted on the other side. Double-sided boards have copper clad on both sides and are often used when the circuit is very complex. The components are mounted so that they will not short out the traces on their side of the board.

The circuit layout is applied to the copper clad in a variety of ways. Commercially made circuit boards are produced by first coating the copper with a light-sensitive chemical layer called the *resist* (or *photoresist*). An exact-size film negative is then placed over the copper clad. The negative contains the actual circuit board layout. An exposure is made by bathing the board in strong ultraviolet light. The light passes through the negative and strikes the sensitized copper underneath.

After exposure, the board is then dipped in a resist developer. What comes out is a copper-clad board where the portions of the copper that were not exposed to light have turned black or a dark gray (this is the positive method; the negative method produce blacks or grays in those areas that were exposed to light).

The board is then dipped in etchant solution. The etchant is a strong acid that eats away at the copper. The black/gray areas ''resist'' the action of the etchant, thereby leaving the circuit pattern on the base of the board.

You can duplicate this photographic method in your own shop, although the process is time consuming and can be expensive. There's more about the photo method, and others, later in this chapter.

Choosing Copper Clad

Materials for making printed circuit boards come in various forms and thicknesses. Copper clad is available either single or double sided. Unless you are making a double-sided board, stick with single-sided clad because you'll waste etchant when making single-sided boards with double-sided clad.

Cladding is applied in various thicknesses, usually 1.0 oz/ft^2 (about 35 μm, or a little less than half the thickness of a human hair). The exact thickness of the clad is not a major consideration with most circuits, except where high voltages or currents are present. Increase the width of the traces if the copper clad is very thin.

The thickness of the board material (called the substrate) also varies. Stay away from thin epoxy boards—they break too easily. A G-10 grade board (standard thickness of 0.062 inch) made of epoxy resin is perfect for the job. Paper-base phenolic and epoxy resin boards (FR-2, FR-3) are flame resistant and more expensive, as are the temperature and flame resistant FR-4 and FR-5 grade boards.

You can buy copper cladding at retail and surplus outlets. Surplus stores usually carry off-size cutoffs left over from printed circuit board manufacturing. While this stuff can be successfully used, it's often filthy and might be heavily oxidized. Be sure to clean it thoroughly, as discussed in the next section. Boards purchased through retail outlets provide you with a choice of cladding and board thicknesses, as well as size. An assortment of sizes ranging from about 2 × 2 inches to 5 × 7 inches should do nicely.

Cutting and Cleaning

The board should only be as large as the circuit layout and no larger. Etching a 5 × 5-inch board that contains only 2 × 3 inches of circuit space wastes etchant solution and costs you money. If the board is too large, cut it with a hacksaw, razor saw, or hobby scroll saw. A scroll saw, such as a Dremel reciprocating hobby saw, is a good choice for all but very heavy copper boards. Draw or scribe a line in the copper to ensure a straight cut. You can smooth the edges of the board with a file or fine-grit sandpaper.

No matter what process you use, be sure that the copper is *completely clean* before applying any chemicals (dirt and grease act as a resist and foul up the etching process). Scrub the copper using a nonmetallic scouring pad and Ajax or similar cleansing powder. Scrub for two or three minutes to remove all grease, dirt, and oxidation. When clean, the water should form a smooth sheet on the copper, not bead up like water on a newly waxed car.

Laying Out the Board

The easiest way to make a printed circuit board is to use a preprinted template, such as those published in electronics magazines like *Radio-Electronics* or *Modern Electronics*. These templates can be photographed, transferred to acetate film, or used as is to produce a printed circuit board. Let's take a closer look at each method.

Photographic method

The photographic method involves making a photo positive or negative of the original artwork and using it as a mask to expose an image on a piece of photosensitized copper clad. The clad is then processed in a developer to bring out the resist pattern.

Have a local printer produce a photo negative or positive of the original artwork that appears in the magazine or book. A negative inverts the polarity of the image; black areas become clear, and white areas become black. A positive retains the polarity of the image: black areas remain black, and white areas become clear. Avoid using a photocopy of the original because this can introduce sizing errors and makes for a blurry negative. The cost of the negative or positive film depends on the size of the artwork, but is usually $5 or less.

Typically, the artwork appearing in magazines and books is printed 1:1; the artwork is the exact size as the finished circuit layout. Occasionally, the artwork will be smaller, in which case you need to tell the printer to enlarge it. If the artwork was reproduced at 50 percent of its normal size, for example, have the printer enlarge the image 200 percent. Note, however, that enlarging a board layout often yields poor results. You might need to enlarge the original onto paper, clean it up with ink and white-out, then make a 1:1 negative or positive.

Be sure to use the right kind of sensitized copper clad or photoresist spray. Most layouts in magazines and books are designed to be used as film positives—the black portions represent areas that will contain resist and the white areas will be washed away in the etchant. If you make a film positive for use as the mask, this requires a positive-acting sensitizer. The blacks turn into resist and the clear portion of the film is etched away. Conversely, a negative-acting sensitizer turns whites into resist and the blacks are etched away.

You need a developer that is compatible with the type of photoresist you are using. When buying photoresist spray, the correct developer is generally packaged with it. Whether you use a positive or negative process, this technique can be used to make any number of printed circuit boards.

Also be sure that the layout is transferred to the board in the proper orientation. Reversing the layout will make a mirror-image of the board, rendering it useless. Spend some time thinking about how the layout will transcribe to the copper clad, and be sure the layout is not reversed.

In a darkened room, spray a coating of photoresist onto a clean, copper-clad board. You can also purchase precoated copper clad where the photoresist has already been applied. This method is easier, but more costly, particularly if you are making many PCBs.

Place the film over the sensitized board, and insert both in a suitable exposure holder. You can make your own exposure holder using the diagram in Fig. 13-2, or use a glass contact-printing holder purchased at a photographic shop.

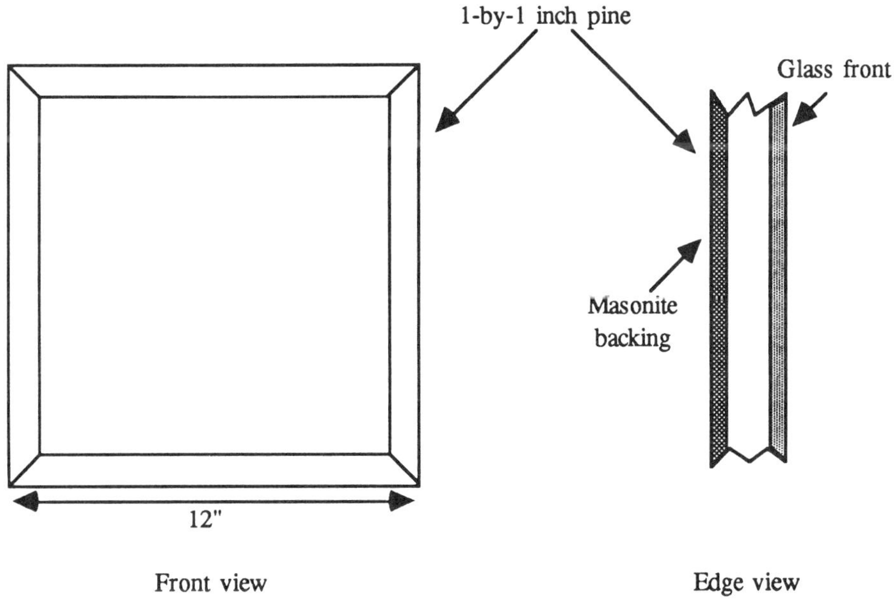

Fig. 13-2. Design of homemade holder. Sandwich the board and artwork between a pane of glass and a wood backing, as shown.

Note the orientation of the film. Make sure that the film isn't reversed, or you might etch the board with a mirror-image of the circuit layout. This will require some thought on your part, so don't rush through it.

It's always a good idea to place the film negative so that the emulsion (the dull side) faces the copper clad. This produces a sharper image on the board. However, you might need to reverse the film (emulsion side facing out) if the original from the magazine or book is also reversed. Some magazines print circuit artwork in reverse (called left reading), to facilitate transferring the image using nonfilm methods. One such method uses mineral oil, as detailed later in this chapter.

One way to make sure you orient the film properly is to look at the component layout diagram that is usually printed with the circuit layout. This diagram shows where each component goes on the board. There might be shaded portions of the component layout diagram that show where the pads and traces are located on the underside of the board.

Now look at the film and orient it to match the shading of the component layout diagram. Most boards are not symmetrical, so it's easy to tell the left from the right side. Place the sensitized board into its exposure holder. Finally, flip the film over (left to right) and place it over the board.

The sensitized board can be exposed in a number of ways. If the sun is out, you can expose it to ultraviolet rays. Exposure time varies between just a few minutes to over 15 minutes. The many makes and brands of photoresist require different exposure times, so check with the instruction sheet.

You can expose the board faster by using an ultraviolet tanning lamp, such as the one shown in Fig. 13-3. Remember that the lamp gives off intense ultraviolet rays, and you should not be exposed to it for long periods. Wear eye protection if necessary.

Once exposed, the board must be developed in a suitable developing solution. Follow the manufacturer's recommended times for developing. After developing, the board is ready to be etched, as detailed in chapter 14.

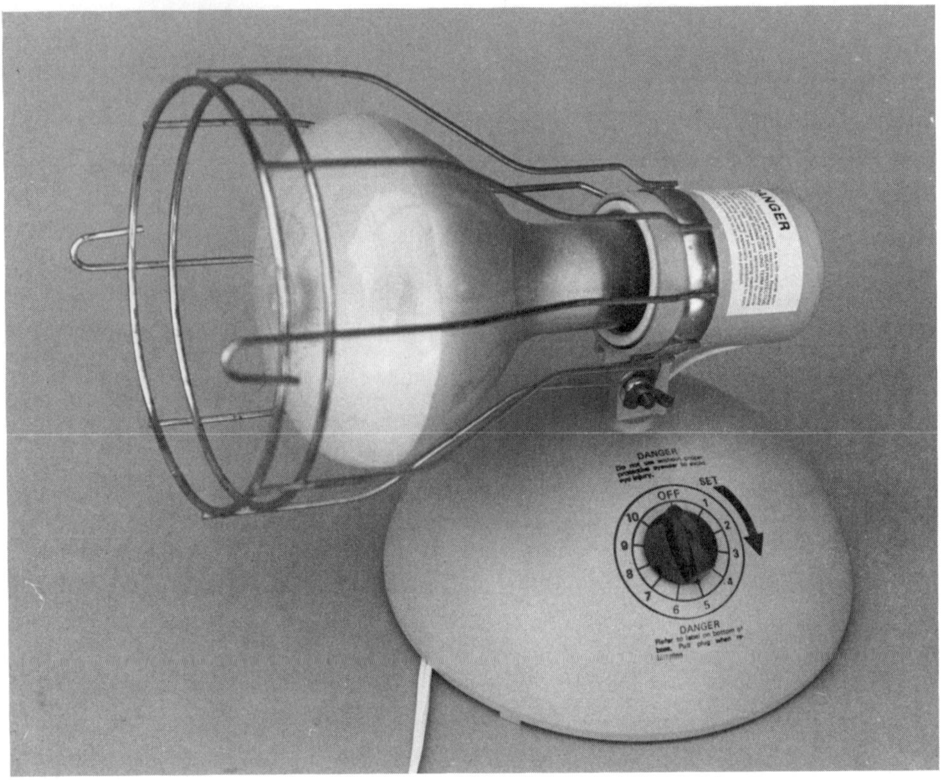

Fig. 13-3. An ultraviolet tanning light.

Transfer film method

The photographic method requires the use of sensitized copper clad, negatives, and a strong exposure to ultraviolet light. All in all, it's a lot of fuss, especially if you're just making one board. If you plan on just making one or two circuit boards from preprinted artwork, you might want to consider the transfer film method.

The transfer film method involves nothing more than a sheet of clear acetate—specifically the type used to print overhead transparencies—and a plain paper copier or laser printer. When the artwork is copied onto the transparency, it fuses black toner onto the sheet. This toner is an effective resist to the etchant solution used to dissolve unwanted copper from the circuit board. You transfer the toner from the transparency to the copper clad using the heat from an iron. While the method might seem backward and anything but high tech, it works quite well.

While overhead transparency acetate is an adequate material as a transfer film medium, you'll get overall better results using transfer film specifically designed for the job. Meadowlake Corp. (see appendix A, "Sources") sells sheets of its TEC-200 film, which is ideally suited for making printed circuit boards. The cost is a little higher than conventional overhead transparency acetate, but it's still reasonable—about 25 cents per 8 × 11-inch sheet.

You need a plain paper copier in top-notch working condition, or the image recorded on the transfer film will be gray and grainy. If you don't have such a copier at your disposal, take your printed circuit board artwork to your local copy center. Many neighborhood copy shops use state-of-the-art high-speed plain-paper copiers that can produce jet-black images on the clear acetate film.

Note that some copiers require you to use acetate sheets with a leading white strip. This strip is used by the copier to sense when a piece of paper is feeding through critical parts of the machine. Without the strip, the acetate could jam. In any case, check with the copy shop and ask if they need a specially marked transparency film.

As with the photographic method, you need to monitor the orientation of the image. Original artwork that has been reversed (left-reading, or mirror image) must be reversed again. You can accomplish this by:

- Use a plain-paper copier that can reverse the image (some of the top models have this feature).
- Make one copy of the artwork onto the transfer film, turn the film over, and use it to make a second copy. Place a piece of white paper behind the transfer film for a clean background.

Once the artwork has been copied to the transfer film, look carefully for washed-out areas or incomplete toner adhesion. With a soft tissue, carefully test the toner to make sure it's firmly bonded to the acetate sheet. Some copiers don't fuse the toner well to overhead transparency film. If this is the case, try a different copier or increase the heat setting in the copier. *Beware:* Do not increase the heat setting beyond that recommended by the manufacturer of the transparency film, or you could end up with a sticky, gooey mess inside the copier!

Protect the image on the transparency by covering it with a blank sheet of paper. Toner that is accidentally scraped off the film will cause a void in the printed circuit.

Now you're ready to transfer the toner from the transparency to the copper clad of the printed circuit board. Clean the board thoroughly using a household cleanser (like Ajax) and a sponge. The copper should be bright and shiny. Be absolutely sure there is no oil or dirt on the board.

Place the PCB with the copper side up on an ironing board. Cut the transfer film to size so that it is no larger than the board. Carefully place the transfer film over the board so that the toner side faces the copper clad. Secure the film into place with strips of masking tape. Do not tape over toner areas. Place a small cheesecloth or several layers of paper towel over the PCB and transfer film.

Set your iron to cotton-linen, or medium-high heat. Let it warm up, then apply the iron over the board. Be careful to avoid wrinkles by keeping the cheesecloth or towel flat. Apply steady and firm pressure for 15 to 20 seconds, as shown in Fig. 13-4. Wait 10 to 15 seconds, then gently lift the cloth and look at the transfer sheet underneath. Carefully peel back a small corner of the film to see if the toner has transferred to the copper clad. If the toner hasn't completely transferred to the copper, replace the film and apply the iron for another 15 to 20 seconds.

Fig. 13-4. Ironing TEC-200 film (shown here without protective cloth).

You'll have to experiment with the exact time for proper toner transfer. It might be as short as 15 seconds or as long as a minute. Do not attempt to speed things up by increasing the temperature setting of the iron. This causes the transparency film to "wilt," causing a distortion of the circuit layout.

After the toner has been transferred, wait for the film to cool (if it's hot to the touch, leave if for a minute or two). Then gently remove the spent film, and discard it—it cannot be used again. Carefully inspect the pattern on the copper clad. The toner should be well adhered, with few, if any, voids. Use a magnifying glass (3× or 4× is fine) to get a close-up view. If voids or skips are apparent, fill them in with a resist pen, available at Radio Shack and most electronics stores. The pen uses a jet-black ink that resists the action of the etching solution.

You might find it necessary to "burnish" the toner into the copper clad. A small wood or rubber roller is ideal for this purpose. I have had good success with a wooden wallpaper seam roller. Be careful to burnish the toner onto the copper with a rolling action, not with a scraping action, or you might scratch off the circuit layout.

The board is now ready for etching, as detailed in chapter 14, "Printed Circuit Board Etching."

Mineral oil method

Printed circuit board artwork that appears in books and magazines can sometimes be successfully transferred using a unique mineral oil method. The mineral oil doesn't transfer the actual ink used in printing the artwork, but it makes the paper semitransparent. The paper, soaked in mineral oil, acts as a type of photograph film negative, and you produce the printed circuit board in much the same way.

The mineral oil method requires that the magazine or book page be printed on one side only. If something else is printed on the opposite side of the artwork, you'll need to make a plain-paper copy of it. Use a copier that is in prime condition: you need a copy that is jet black and without streaks or specks. You might also want to use a copy of the original if you don't want (or can't) disturb the magazine or book.

Cut the artwork out of the magazine or book and trim it so that it will fit onto your copper-clad board. If you are using a presensitized board, there's no need to clean it. Otherwise, follow the steps provided earlier in this chapter for cleaning copper clad. Once clean and dry, spray the photoresist onto the board.

In a darkened room, carefully tape the paper onto the copper clad. Be sure the paper is smoothed out and that no tape covers any area of the artwork. With a soft brush, coat the paper with mineral oil. Let the oil soak into the paper so that the paper becomes semitransparent. You should be able to see the black printing against a faint background of the copper clad.

As with the photographic and transfer film methods, note the proper orientation of the paper. The mineral oil technique works best when the printed side of the paper faces the copper clad. This requires that the original artwork be printed in the book or magazine in left-reading, or mirror-image orientation. If the artwork is printed right reading (the layout appears on the page the same way it will appear on the foil side of the printed circuit board), you'll need to turn the paper over.

The board is now ready for exposure. Mount the board in an exposure frame, as detailed earlier in this chapter. Because the paper is held securely to the copper clad with tape and the wetness of the mineral oil, there's no need to use a sheet of glass in the exposure frame.

Exposure times are between three and 10 times longer than when making an exposure using ordinary photographic film. Read the instructions that came with the presensitized board (or photoresist spray) and experiment until you find the correct exposure. Note that ultraviolet tanning lights provide the fastest exposures, but these are expensive (about $100 for the fixture and bulb) and they can be dangerous to use. Always avoid prolonged exposure to the light and shield your eyes at all times.

After exposure, remove the oil-soaked paper and rinse off the board in cold running water. Process the board in the developer liquid as recommended by the manufacturer. After the board dries, it is ready for etching. See chapter 14, "Printed Circuit Board Etching," for more details.

Making Your Own Layout Artwork

You can also make a printed circuit board completely from scratch, using one of two popular methods:

- Apply the resist directly to the copper clad. Dry-transfer direct-etch kits (obtain from Datak or Radio Shack), provide pads for ICs and other components, as well as spools of thin black tape for the traces. You apply the resist in the same manner as making signs using dry transfer lettering. The direct-etch method is practical for designs where you want to make just one board.
- Draw the layout using dry-transfer sheets, computer, plotter, ink pen, or other method, and use the drawing as a master for making one or more PCBs. The master is a film negative or positive for exposing the surface of the sensitized copper clad. You can buy presensitized boards, or apply the sensitizer with a brush or spray can. Exposure typically requires a strong shortwave ultraviolet light source (not a "black light").

You can also draw the resist directly onto the copper clad using a resist pen. A black felt tip pen will work, but a pen made especially for printed circuit board etching is the best choice. However, this method lacks the precision necessary when working with miniature components such as ICs and transistors.

Using direct etch

If you've ever used dry-transfer lettering, then you can easily master the direct-etch method of making printed circuit boards. To make a direct-etch printed circuit board:

- Draw the layout on a piece of paper. You can draw the pads to scale by using a plastic template, available at graphics arts stores. While drawing, think of the best location for components and the best way to route traces between the pads. You might need to move pads and traces; it's easy to do at this point because you're only working with pencil and paper. Graph paper with 1/10-inch graduations can be helpful when laying out the circuit.
- Clean the board following the instructions provided earlier in this chapter.
- Let the board dry completely. Set it out in the sun if necessary. *No* moisture can be present.
- Using your paper layout as a guide, use the transfer sheets to place the pads for the first component at the proper locations on the copper clad. The transfer sheets include single pads, as well as multiple pads for ICs and transistors, as shown in Fig. 13-5. The multiple pads help you maintain scale and spacing.
- After transferring a pad to the board, burnish onto the copper clad using a burnishing tool or other hard plastic instrument, like the cap to a Bic ballpoint pen.
- Repeat the process for several more components.
- Apply the traces between the pads using the straight and curved lines provided on the dry-transfer sheets or use transfer tape. Burnish the traces onto the copper clad so that they won't come off.
- Continue adding pads and traces until the entire circuit board is complete.

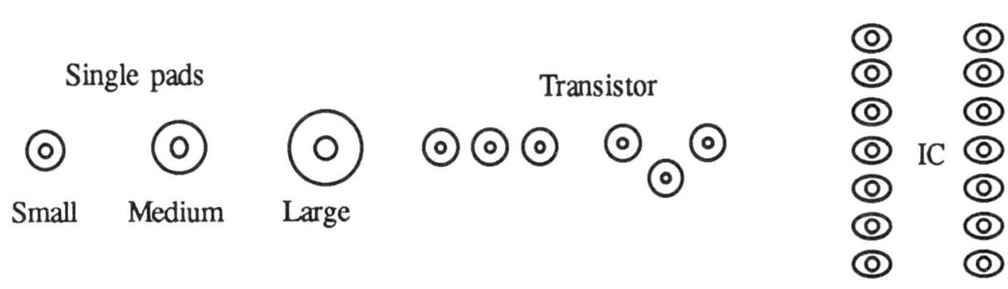

Fig. 13-5. An assortment of rub-on pads.

Give the board one final burnishing. Look for errant pads and traces that might have accidentally come off the transfer sheets; remove these with a razor blade. The board is now ready for etching, as detailed in the next chapter.

Creating an artwork master

The direct-etch method is ideal if you want to make just one circuit board. But if you want to create several circuit boards from one design, you'll want to think about creating an artwork master. The master can then be used to make one or more printed circuit boards using the photograph or transfer film methods discussed earlier in this chapter.

There are two popular approaches to creating artwork masters for printed circuit boards. The first is that you can draw the master by hand onto a piece of drafting vellum. Most better-stocked graphics arts stores sell a wide range of tools and supplies for designing printed circuit boards including rub-on pads, traces, and tracing templates. You're better off drawing the circuit at 200 percent and reducing it 50 percent to make a photographic positive or negative.

An alternative is to draw the artwork using a computer, such as an IBM PC or Apple Macintosh. You can use special applications programs designed for the task (there are several for both of these computers), or you can use a drawing/drafting program like MacDraw (for the Macintosh) or AutoCAD (for the IBM PC). The artwork is best when printed by a plotter or laser printer. For best results, create and print the artwork at 200 percent or more, and then have it photographically reduced to make a film positive or negative. Figure 13-6 shows a sample circuit board layout created entirely by the Apple Macintosh and printed on a laser printer.

Making Double-Sided Boards

Double-sided boards have a foil pattern on both sides. They are used mostly with computer circuits because of the high number of circuit lines stretching from one IC to the next. Many of the projects you will build, including those in this and other books and magazine articles, require the use of single-sided PCBs. But a special design of your own or from an electronics publication might require double-sided construction.

Double-sided boards are produced in the same way as single-sided ones, except you must maintain perfect registration between the two sides. The pads for ICs and other components must match exactly, or a short circuit could result. Except for the photographic methods, most other means of creating printed circuit boards are not reliable enough to create a double-sided PCB.

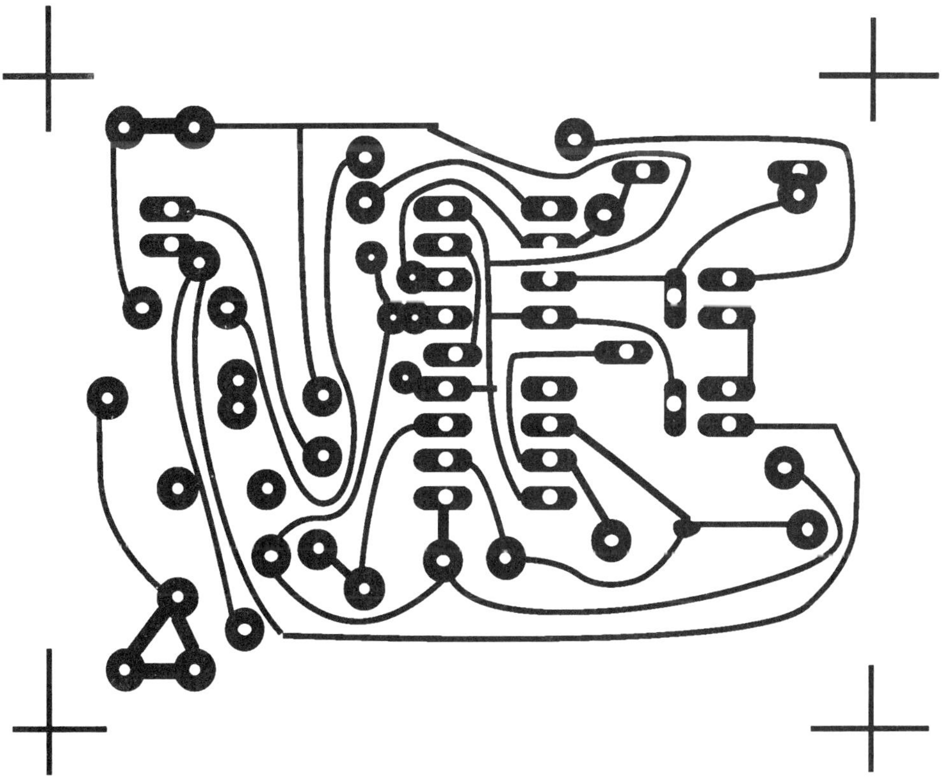

Fig. 13-6. A circuit board drawn by computer, and printed on a laser printer. I used an Apple Macintosh, MacDraw drawing software, and a Laserwriter laser printer.

The holes for component mounting in commercially produced double-sided printed circuit boards are "plated through" to ensure proper contact. This plating process is beyond the means of most hobbyists. Rather, you will need to solder your components to the pads on both sides of the board to ensure the components make electrical contact with the circuits on both sides.

14

Printed circuit board etching

IN CHAPTER 13, YOU LEARNED HOW TO CREATE A RESIST PATTERN ON A NEW sheet of printed circuit board material. The next step is etching that board to remove the unwanted copper. The copper that remains forms the printed circuit that enables your project to function.

This chapter details the steps in printed circuit board etching. You'll learn how to mix the etchant fluid and how to prevent possible injury. You'll learn how to speed up the etching process by using hot etchant fluid or by adding air bubbles with an aquarium pump. Also included in this chapter are instructions for drilling component holes.

Inspect the Board

Before etching the board, inspect if for errors in layout, missing traces and pads, skips in the resist pattern, and other gremlins that can cause a poor job. If you created the resist pattern from artwork appearing in a book or magazine, compare your board with the printed layout. Follow the traces from pad to pad and note any discrepancies. If you created the resist pattern from your own design or by using the direct-etch method, carefully review your work and compare it against a schematic or paper drawing. Be sure that pads and traces do not come too close together. At a minimum, all pads and traces should be $1/32$ of an inch apart, preferably more.

The traces and pads on a direct-etch board should be firmly burnished into the copper clad. If the resist is loose, etchant fluid will leak underneath and undercut the copper pattern. You'll end up with voids and an incomplete circuit. Repairing the board after a shoddy etching job—if it can be done at all—is time consuming and frustrating.

After inspection, wet a cotton ball with isopropyl alcohol and gently clean the exposed parts of the board. Don't apply too much alcohol because some types of resist are susceptible to it and could melt or distort. Let the alcohol dry completely before immersing the board in the etchant fluid.

About Printed Circuit Board Etchant

Most circuit board etchants, whether liquid or powder, are toxic and highly caustic. The etchant chemical should never be allowed to come into contact with your skin or clothes. If you do get some etchant on your fingers or hands, wash it off immediately. Unless the etchant is extremely concentrated, it will not burn your skin if you remember to wash it off immediately.

Etchant does stain skin and clothing. Wear a smock while etching your circuit boards and be sure to wear old pants and shoes. Wear protective goggles, like those used in wood and metal shops, to prevent injury to the eyes should etchant splash into your face.

All etchant solutions give off caustic fumes, which can severely injure the mucous membranes in your nose and throat. The effect is not often instantly noticeable. You may etch one or two boards and not be aware of the fumes entering your nose and mouth. But an hour or two later, you'll feel an intense burning in your nose or throat that can last several days. Prolonged exposure to etching solution fumes could cause serious bodily injury, so be sure to etch your circuit boards only in a well-ventilated area.

You can wear gloves to protect your hands against burns and stains. You'll need gloves that provide a good amount of manual dexterity. Disposable plastic or latex gloves are a good choice; cotton or polyester gardening gloves are not.

Mixing the Etchant Solution

Etchant is available in three popular forms:

- Liquid, not concentrated
- Liquid, concentrated
- Powder

Liquid, unconcentrated etchant is available at Radio Shack and most electronics stores. It comes in a plastic bottle ready for use. Just open the bottle, pour the etchant solution into a plastic (*never* metal) tray, and you're ready to go.

Unconcentrated liquid etchant can be used to make more than one board, depending on the size of the board. The etching action is reduced as the surface area of the board increases. If the board measures 4 × 6 inches with one side, there are 24 square inches of copper clad.

Note the recommended usage on the etchant bottle. Your particular solution might be able to etch up to 50 square inches of copper clad, assuming you use the entire contents of the bottle. If you use less, the expected amount of coverage is reduced. When in doubt, toss the used etchant fluid. Using weak etchant can cause overly long etching times, which can undercut the resist and lead to voids in the copper pattern.

Concentrated liquid etchant must be diluted before use. For best results, dilute the etchant with hot water to increase the etching action. Typical dilution ratios are 2:1, 3:1, and 4:1. The higher the ratio, the longer the concentrate will last. For best results, though, balance the thrifty use of the concentrate with etching times. The weaker the etchant, the longer it will take to remove the excess copper.

Powder etchant must be mixed before use. One packet of powder etchant generally makes one or two quarts of unconcentrated etchant. You can mix the powder to make a smaller amount of liquid, then dilute the mixture when you are ready to use it.

About Etching Trays and Tongs

No matter what type of etching solution you use, always pour it into plastic containers. Avoid all metals because the etchant will react with it. Be sure that the stopper or cap for your containers are not metal nor have any metal parts inside.

When using powder or concentrated etchant, mix with warm water. While stirring, chemical reaction will heat up the etchant even more, so be sure to use a plastic tray that can be used in *very* hot water. Processing trays for photography generally make ideal etching troughs. For safety, check the tray before use by filling it with hot (150 to 180 degrees) water. The tray should not become soft or melt!

The design of the tray is only secondary to the material used in it. For best results, the tray should have ripples or ridges on the bottom. These ridges allow the etchant solution to flow freely under the board while it stews in the tray. Most photographic processing trays have ridges. If yours doesn't, add a series of raised mounds using ordinary 5-minute epoxy, as shown in Fig. 14-1. The mounds should be no more than about 1/16 inch high, spaced about an inch apart. These mounds keep the circuit board from sticking to the bottom of the tray. Let the epoxy dry and cure thoroughly, which normally takes several hours.

In addition to a plastic tray large enough for the circuit board you are using, you'll need two plastic or bamboo tongs to handle the board. These tongs are available at any photographic store.

Etching the Board

To use the etchant, pour it into the plastic tray carefully. Avoid spills and splashes. Pour enough etchant to provide a pool at least 1/8 inch thick, preferably

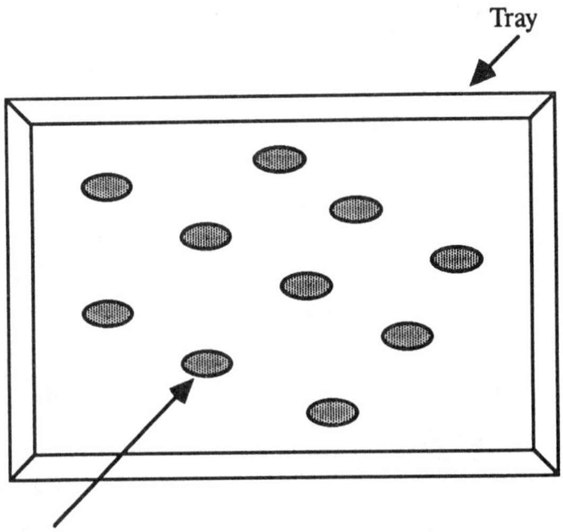

Fig. 14-1. Apply beads of epoxy to the bottom of the etching tray.

1/4 inch. Dunk the board into the tray and constantly rock it back and forth to agitate the liquid. Keep the board in the soup for 10 to 30 minutes (depending on the type and strength of the etchant) or until all the excess copper is removed. Use the plastic or wooden tongs to lift the board out of the tray from time to time to check its progress.

Notice that the copper comes off first at the edges and areas closest to the resist. Large, open areas of copper are stubborn and take two to three times as long to etch completely. You might want to apply extra agitation at those areas of the copper that are not responding as quickly to the etchant. However, be sure that you do not overagitate, as this can undercut the copper under the resist.

As mentioned earlier, the size and number of boards you make determines how long the etchant lasts before it is exhausted. Weaker etchants must be thrown out after they are used to make just one 2 × 3 inch board; stronger etchants can be used to make several large boards. Dispose of the etchant by dumping it down the toilet or pouring it into the sink. Flush thoroughly with water to clean the sink or toilet and make sure that no etchant residue remains. Etchant left on porcelain can stain, making you very unpopular around your house.

Diluted etchant solution used for hobbyist applications usually does not pose a serious threat to plumbing, but the etchant is a pollutant and a toxin. If at all possible, you should take the exhausted etchant to a licensed recycler or approved chemical-waste-disposal site. If you live near an electronics manufacturer, you might be able to interest them in disposing of it for you. Copper can be reclaimed from exhausted etchant solution, so the company might not charge you for this service because they benefit from the recovery of the copper.

Using Hot Etchant

You can speed up the etching process by using a hot etchant solution. If you're using powder etchant and you just mixed it, it's already hot and doesn't need to be heated.

Liquid etchants can be heated by resting the plastic container on a small wire rack placed at the bottom of a 2-quart sauce pan. Fill the pan with water and apply medium heat for 3 to 5 minutes. Don't let the water come to a rolling boil; that could cause the plastic container to melt, and you'll end up with etchant oozing all over the place. The etchant is warm enough when the water is hot to the touch but does not scald you. Remove the pan from the heat and leave the etchant container in place for several more minutes. Periodically remove the container and shake it.

Using Air Agitation

In addition to using hot etchant, you can automate and speed up the etching process by using an ordinary aquarium air pump to make an "air burst agitator." Connect the pump to a length of plastic tubing and poke holes, using a hot needle, into the tubing. Lay the tubing along the bottom of the tank, as shown in Fig. 14-2, and cement into place using Weldit or some other all-purpose plastic glue. Do not use airplane model cement. Plug the end of the tube so that air won't escape there.

Connect the other end of the tube to an ordinary aquarium pump. For safety, add a check valve near the pump outlet. The valve will prevent etchant fluid from siphoning into the pump. You can also prevent siphoning by keeping

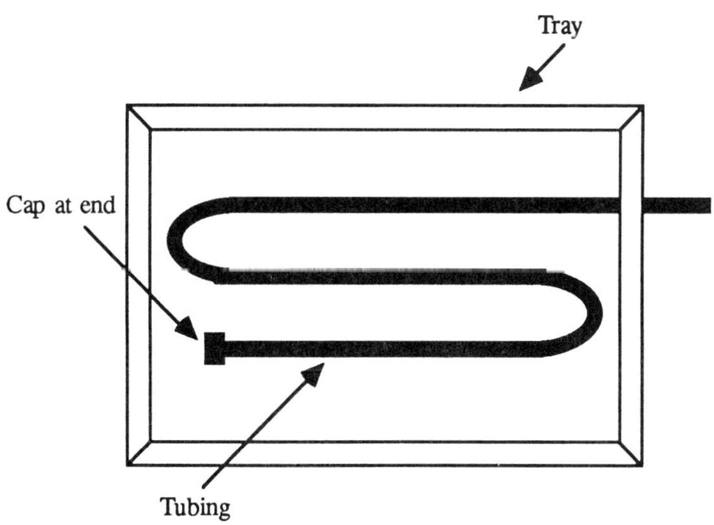

Fig. 14-2. Air tube glued to the bottom of the tray.

the pump at a higher elevation than the etchant tray. A diagram of the air-burst agitator, with pump, tubing, check valve, and tray, is shown in Fig. 14-3.

When the pump is on, air is forced out of the holes along the length of the tube and causes bubbles in the etchant. These bubbles act to agitate the etchant and keep it moving in the tray. The disadvantage of this method is that you need to fill the tray with more etchant.

If you are preparing many circuit boards at once, you can build a small production etching tray using the air-burst method. Build a tray large enough to accommodate several boards stacked on end. For most applications, the tray should measure 8 inches high by 10 inches long by 5 inches wide. This will hold about six or seven boards, each measuring about 7 × 9 inches. Construct a frame to hold the boards during etching. You can also use *plastic* sheet film holders, available at many photographic shops (remember: do not use metal sheet film holders).

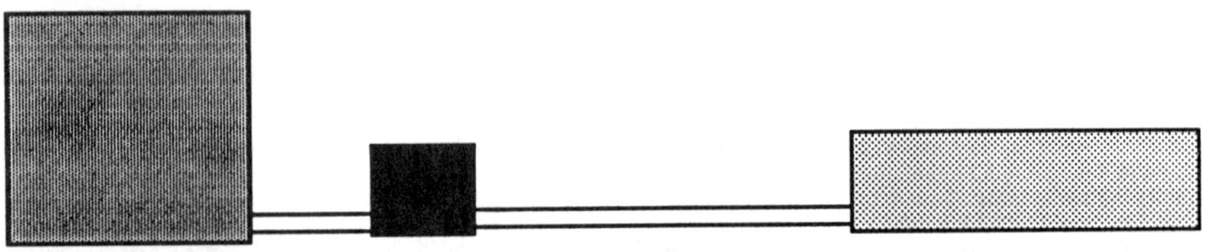

Fig. 14-3. Air-burst etching setup diagram.

Final Prep and Drilling

Etching is complete when no traces of exposed copper can be seen. The black resist remains for the traces and component pads. Assuming that the etching process was done correctly, the copper should remain under the resist.

After etching, rinse the board under cold running water for 15 to 20 seconds. Be sure to rinse the etchant from the back side of the board as well.

Remove the resist with lacquer thinner or a thorough scrubbing with a nonmetallic scouring pad and cleaner. I regularly use Ajax cleanser and the green Scotchbrite scouring pads. When the resist is completely removed, the copper should be bright and shiny, with no evidence of undercutting. The undercutting is when etchant has oozed under the resist and has attacked the copper you don't want removed.

If you find that the board was overetched and some of the traces are missing, you can repair the board by soldering short lengths of wire to bridge the gaps, or apply copper tape to the missing portions. A variety of copper tape and pads are available for making or repairing PCBs. One of the most popular is EZ Circuit, sold by Bishop Graphics (see appendix A). The copper pieces have adhesive backing for fixing them onto the board.

Drill the board using a small 0.040 or 0.070 drill—the smaller drill is recommended for IC sockets and small resistors and capacitors. A drill press or drilling stand is recommended. For best results, keep most of the bit tucked inside the drill motor chuck. With only 1/4 to 1/2 inch protruding from the chuck, there is less chance of breaking or bending the bit.

You will need to drill from the foil (copper) side of the board to the back. Prevent splitting and chipping the back of the board by placing a sheet of scrap wood underneath it. Do not use particle board; it will quickly dull the drill bit.

Position the drill exactly within the hole on each pad. If a component pad lacks a hole, drill in the approximate center. Most components do not need precision drilling, but some—most notably ICs—require that you drill the hole within about 1/50 of an inch of the proper spot. If the drill bit ''dances'' around the copper foil before it digs in, use a center punch to make a small dimple in the board to help you aim the bit at the exact spot for the hole.

After drilling, inspect the copper for burrs, and remove them with an emory board or fine steel wool. Examine the back of the board for chips and cracks. Remove broken pieces of epoxy and file away the rough spots. Thoroughly wash the board to remove pieces of copper, epoxy, wood, steel wool, and other contaminants. Use a nonmetal scouring pad (like 3M Scotchbrite) and cleanser to thoroughly clean the copper foil. You are now ready to solder the components to your custom-made printed circuit board. If you are not going to use the board for a while, place it in a plastic baggie and store it in a safe place where it won't get dirty.

15

Wire wrapping

MAKING A PRINTED CIRCUIT BOARD FOR A ONE-SHOT APPLICATION IS TIME consuming, as detailed in the last two chapters, though it can be done with the proper kits and supplies. Conventional point-to-point solder wiring is not an acceptable approach when constructing digital and high-gain analog circuits, which represent the lion's share of electronics you'll be building.

An alternative is *wire-wrapping*. Wire-wrapping is a point-to-point wiring system that uses a special tool and extra-fine 28- or 30-gauge wrapping wire. When done properly, wire-wrapped circuits are as sturdy as soldered circuits, and you have the added benefit of making modifications and corrections without the hassle of desoldering and resoldering.

This chapter details the art and science of wire wrapping. You'll learn about materials used in wire wrapping, how to ensure a solid wire-wrapped joint, how to "daisy-chain" wire-wrap connections, and more.

Wire-Wrap Basics

When working only with low-voltage dc (not ac or high voltage), electrical components can be mounted on a perf board and connected using wire-wrapping posts and wire. No soldering is involved; you just wrap the wire around the posts with a tool. One big advantage of wire wrapping is that changes are relatively easy to make. Simply unwrap the wire and reroute it to another post.

Wire wrapping is used most in IC-intensive circuits. Each IC is mounted in a wire-wrap socket, which is cemented or soldered in place on the board. The sockets use extra-long posts that can accommodate up to about five wrapped wires (four is the practical maximum). Each wire is wrapped around the post like the figures on a totem pole. Most designers try to limit the number of wires on each post to two or three in case changes must be made to the circuit. Once the wire is wrapped six to eight times around the post, the connection is solid and is as secure as a soldered joint. Figure 15-1 shows a wire wrapped around a wire-wrapping post.

Fig. 15-1. Wire wrapped around post.

Wire-Wrapping Tools and Supplies

The basic ingredients for wire wrapping include:

- Perf board
- Sockets for ICs and other components
- Posts for tie points
- Wire
- Wrapping tool

Let's take a closer look at these tools and supplies.

Perf board

You can use almost any perf board where the holes are spaced the proper distance for integrated circuits; that is, the holes are spaced 1/10 of an inch apart.

Perf boards with solder pads are best, but you can use almost any style of perf board you wish. When using perf boards that lack solder pads, you need to glue the sockets and posts into place to prevent them from coming loose.

Sockets and posts

Wire wrapping requires the use of special sockets for mounting ICs and discrete components. These sockets use 5/8-inch-long square pins designed expressly for gripping wire-wrapping wire. A selection of wire-wrap sockets is shown in Fig. 15-2. The socket should be soldered into place (if the perf board uses solder pads) or cemented with gap-filling cyanoacrylate or epoxy glue. Avoid glues that are susceptible to heat or moisture or contain chemicals that will melt the sockets or perf board. This includes ordinary white glue and airplane modeler's cement.

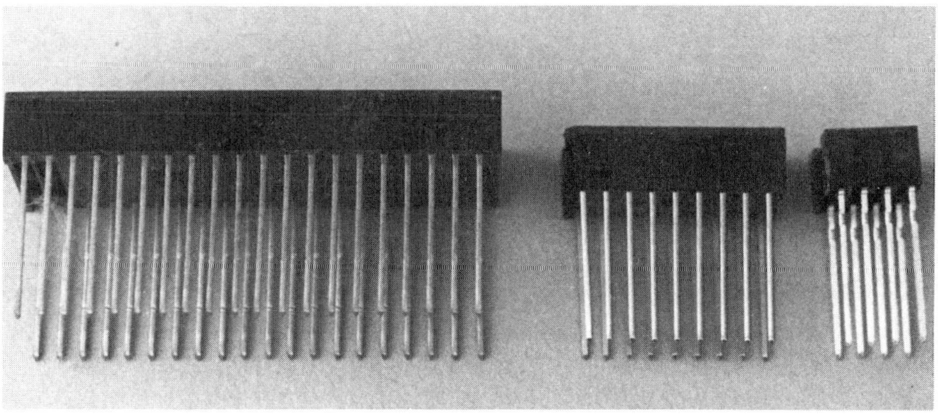

Fig. 15-2. Selection of wire-wrapping sockets. Note the long leads.

You can also purchase individual posts for wire wrapping. These are most often used as common tie points. The posts are inserted into a hole in the perf board with a pair of pliers or a post insertion tool. The insertion tool is by far the easier to use, and its $6 to $8 price is well worth it. Look for this tool at most any well-stocked electronics supply store. The post should be soldered or glued so that it is anchored to the board.

Note that wire-wrap posts are square shaped to firmly grip the wire. The corners dig into the relatively soft metal of the wire to help hold it in place. You can wrap the wire onto rounded posts, but to make the connection solid and permanent, add a dab of solder to the wire. You can use this method to directly connect wires to discrete components such as resistors or capacitors.

A better approach is to cement wire-wrap IC sockets to the board and insert the components into the sockets, as depicted in Fig. 15-3. Bend and cut the leads so that they fit into the socket. If the component is large or wide, use a

Fig. 15-3. Components in sockets.

24-, 28-, or 40-pin socket. Special component sockets with solder terminals are also available.

Wire

Wire-wrap wire comes in various lengths and gauges. For most applications, you want 30-gauge wire in either long spools or precut/prestripped packages (see Fig. 15-4). When using the spools, cut the wire to length, then strip off the insulation using the stripper attached to the wrapping tool (a regular wire stripper does a poor job). The work is already done for you when using the precut/prestripped packages.

Buy a selection of different lengths and always try to use the shortest length possible. When first starting out with wire wrapping, you'll most likely cut the wire either too long or too short. But with practice, you'll be able to gauge the approximate amount of wire you need to form the connection.

Remember that the stripped ends of the wire must be added to the length of the wire you need. For example, if the distance between contact points A and B is 4 inches, you need a wire 6 inches long. In fact, you need about 1/4 to 1/2 inch extra for slack. Avoid forcing a connection with wires that are too short, as that can weaken the wires and bend the posts. The 6-inch length assumes that 1 inch

Fig. 15-4. Sample of precut wire-wrapping wire.

of insulation is stripped off both ends of the wire, which is more-or-less standard. Precut/prestripped can be expensive ($5 or more for a canister of 200 pieces), but it will save a great deal of time and effort.

Wrapping tool

A manual wire-wrapping tool is shown in Fig. 15-5 and is available at most electronics stores. You insert one end of the stripped wire into a slot in the tool, and place the tool over a wrapping post. Give the tool five to ten twirls, and the connection is complete. To remove the wire, use the other end of the tool and undo the wrapping. NOTE: not all wire-wrapping tools have an unwrapping end. These should be avoided because they require you to buy a separate tool for unwrapping the wire.

When you get the hang of manual wire wrapping, you can try one of the motorized tools that are available. Some even allow you to use a continuous spool of wire without the hassle of cutting and stripping. The cost of these tools is high (over $75), so they are not really recommended for beginners.

How to Wire Wrap

Successful wire wrapping takes practice. Before you build your first circuit using wire-wrapping techniques, try your hand on a scrap socket and board. Visually

Fig. 15-5. Wire wrapping in action.

inspect the wrapped connections and look for loose coils, broken wires, and excessive uninsulated wire at the base of the post.

As mentioned previously, most wire-wrap tools are designed so that one end is used for wrapping wire and the other end is used for unwrapping. Undo a connection by inverting the tool and try again. Motorized wire-wrapping tools typically lack a facility for unwrapping the wire, so you'll need to do it with a manual tool.

Note that wire removed from a wire-wrap post will be curled, the same as hair is curled when it has been tightly wound around a hair curler. Replace these wires completely instead of attempting to reuse them. A wire that has been wrapped and unwrapped around posts too many times becomes weak and brittle.

A tool such as the one shown in Fig. 15-6 can be used to route the wire around posts. After making a wire-wrap connection, use the tool to steer the wire around the posts. Wires leading from one IC to another should be routed along the same path. This not only makes the wire-wrap circuit look more presentable, but it aids in double-checking and debugging your work. The chiseled end can be used to clip the loose ends of the wound wire, as depicted in Fig. 15-7.

Most pros use different colors of wire to denote the type of signal the wires

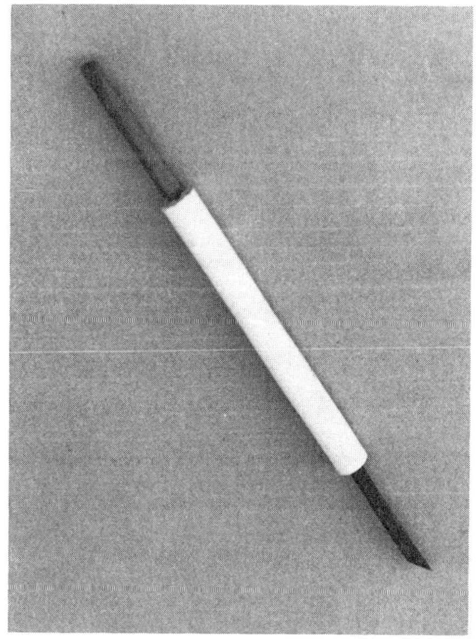

Fig. 15-6. Chisel tool

Fig. 15-7. Using the chisel tool.

carry. You might try the following color-coding convention (this is by no means standard, so you are free to modify it as you wish):

Wire Color	Application
Red	Positive supply
Black	Ground
Blue	Negative supply
Green	Digital signal
Yellow	Analog signal

You can also use color coding to denote different stages of a circuit—yellow for preamplified audio, green for amplified audio, and blue for video. Whatever color-coding scheme you use, keep it consistent and write it down in a notebook or on a label that you can affix to the board.

IC number tags, as shown in Fig. 15-8, can be used to facilitate wiring integrated circuits. The tag slips over the posts of a wire-wrap socket to help you identify the correct pin number of the IC. Because you are working with the board upside down, it's easy to get the pin numbers crossed. With most ICs, a mistake like attaching a wire to pin 14 instead of pin 1 will damage the chip. Unfortunately, once the tag is slipped over the pins and the wire wrapped around the post, it is permanent and cannot be reused. The tags are available in sets of five or ten in all the popular socket sizes: 8, 16, 18, 20, 24, and 40.

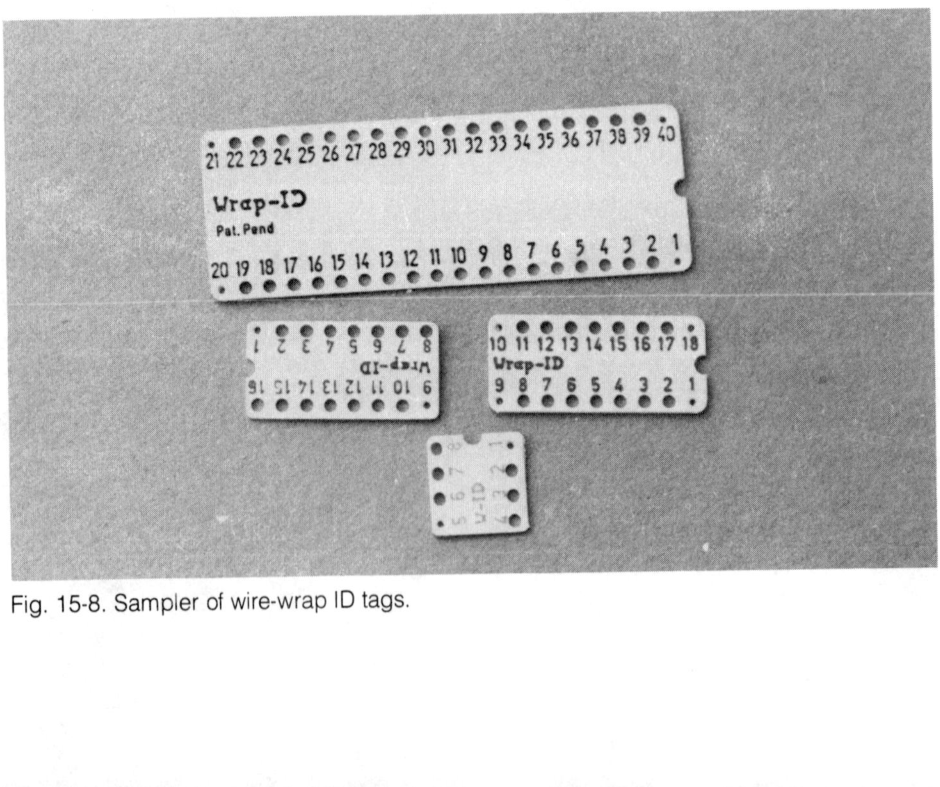

Fig. 15-8. Sampler of wire-wrap ID tags.

Wire-Wrap Chaining

Often, particularly in digital circuits, the wire connections loop from post to post in *daisy-chain* fashion. For example, the wire might start at post A, then go to post B, followed by post C, then terminate at post D.

Normal wire-wrapping procedures can be used to make daisy-chained connections, but it is time consuming. An easier approach is using a slit and-wrap tool, such as the one shown in Fig. 15-9. The slit-and-wrap tool uses specially insulated wire-wrapping wire and is expressly designed for daisy chaining. To use, start the wire on one post, then wind the wire in the usual manner. You then move the tool over to the second post, twirl it 6 to 8 times, and move to the next post. You can repeat this process as many times as you wish.

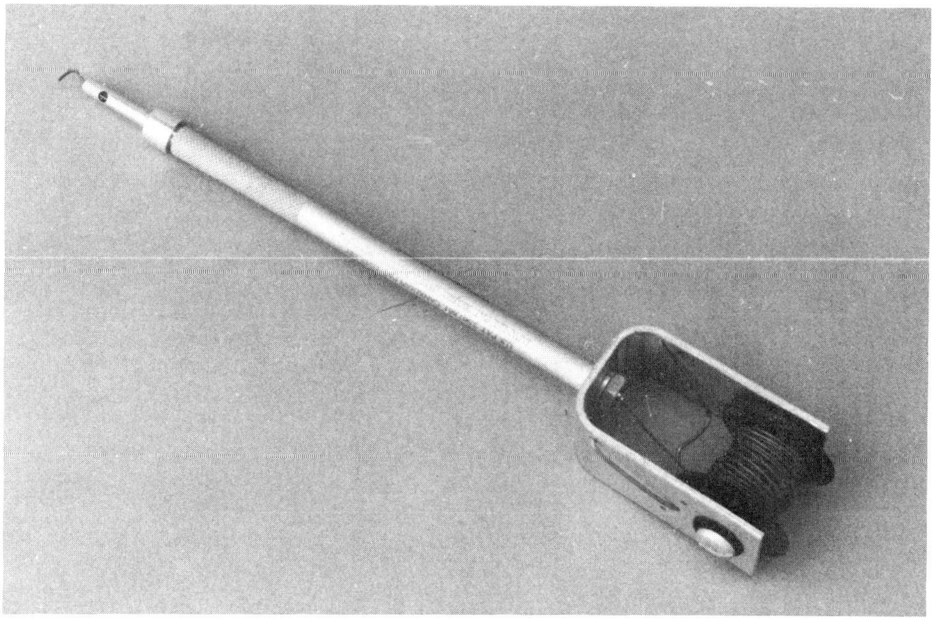

Fig. 15-9. Slit-and-wrap tool.

As with regular wire wrapping, working with the slit-and-wrap tool requires patience and practice. The biggest problem with this method is that some connections might not be secure—the wire might not sufficiently contact the post to make an electrical connection. When using a slit-and-wrap tool, always keep a volt-ohmmeter or continuity tester handy, and periodically check your connections to make sure the circuit is complete. Touch the meter probes to the ends of the posts. A reading (or tone) indicates proper continuity.

Limitations of Wire Wrapping

Your first attempts at wire wrapping will likely end in frustration. Keep with it and you'll get the hang of it. As you gain experience, you might find that wire-wrapping becomes your preferred method of building electronic circuits (it certainly is with me). However, be warned that wire wrapping cannot be used with certain types of circuits. *Do not* use wire wrapping on circuits that:

- create or pass high voltage. As a rule of thumb, limit your wire-wrap circuits to 25 volts ac and 100 volts dc.
- run on or control 117 Vac current. The thin wire-wrapping wire cannot handle the current demand.
- depend on exact values for inductors and/or capacitors. The loose wires and numerous posts of wire-wrapping sockets will interact with the inductance and capacitance of your circuit and can play havoc with its operation.
- use RF, such as radio transmitters or receivers. The wire-wrap posts act as antennas and will render your circuit useless.

Wire-wrapping might seem like a second-best approach to circuit building compared to soldering. But when done properly, a wire-wrapped connection can be as solid as a soldered connection, or even more so. However, there are limitations to the stability of wire-wrapped connections. Wire wrapping should not be used if the circuit will be exposed to caustic elements including smoke, vapors, and other contaminants. Wire wrapping should also not be used if the circuit will be subjected to vibration.

In both of these cases, a soldered, printed circuit board is preferred. If wire wrapping *must* be used, secure the wire-wrap connections by applying a small dab of solder. Once soldered, the connections can be protected with a varnish-type spray. Spray several light coats, allowing the varnish to dry completely between coats.

16

Repairing electronic circuit boards

IN THE OLD DAYS, RADIOS AND TELEVISIONS WERE BUILT USING TUBES. THE tubes acted as modern-day transistors, amplifying and switching signals. They suffered three major disadvantages: they were bulky, they gave off considerable amounts of heat, and they were prone to failure.

Nowadays in our solid-state society, tubes are all but gone from TVs, radios, and electronic gadgets, including the circuit boards that you may build as part of your electronics hobby. Because these circuits are solid-state (no tubes), they live a longer, more productive life. The average transistor radio will last decades; buy one when you're a tot and it will likely outlive you.

When breakdowns do occur in an electronic circuit, they are generally due to accidental damage, like dropping it on the floor or exposing it to excessive heat or moisture. Many circuit malfunctions can be remedied by you, whether the defect is a failed component or accidental damage. The fix is generally no more than reworking a few solder joints or replacing a bad component.

This chapter deals with steps in repairing electronic circuit boards. The chapter doesn't address troubleshooting techniques or using your meter or scope to locate the problem, but just how to fix it.

Resoldering

A common malfunction in an electronic circuit is a faulty solder joint. There are three major types of faulty solder joints:

- Bridge
- Cold
- Incomplete

Let's take a closer look at each.

Bridges

A bridge is caused by too much solder melding together wires or components that should not be connected (see Fig. 16-1). The solder bridge doesn't have to be a huge glob to be damaging: some bridges are so slight that the solder just misses an adjacent component or trace. Movement of the circuit board, or even expansion of the board by heat, can complete the bridge and short the components.

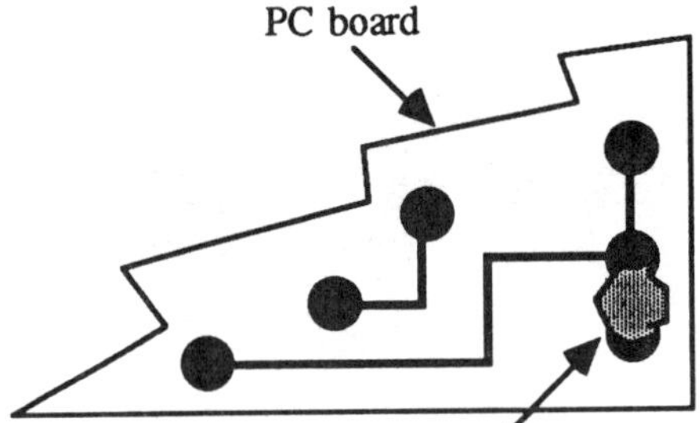

Fig. 16-1. A solder bridge.

Solder bridges are corrected by removing the solder on the circuit board pads, as detailed in the section on desoldering in chapter 11. Once the solder has been removed, new solder can be applied, being careful to avoid a new bridge.

It is important that you first remove the old, faulty solder joint before attempting to fix the bridge. If you merely reheat the solder so that it flows, hoping to separate the blob onto smaller packets, you'll probably end up with a cold solder joint (see the next section). The exception is when the bridge is caused by a very small speck of solder, in which case you can scrape it off with a sharp tool, such as a dentist's pick. Be sure that you do not harm the circuit board.

Solder bridges are sometimes hard to spot. You'll need a magnifiying glass (about 6× will do) to bring the solder job into close view. You can also find solder bridges by using a volt-ohmmeter or continuity tester. If there is continuity between two components or traces (like the positive and negative rails of the circuit board) where there should not be, then a short probably exists, caused by a solder bridge.

Cold joint

A cold joint is one where the solder was not allowed to flow properly over the component lead, wire, and circuit board pad. Cold joints can be caused by:

- Applying the soldering iron to the solder and not to the work.
- Moving the work while the solder is cooling.
- Using an iron that is too cool. A hotter iron is required when soldering large connections.

A cold solder joint has an unusual, dull, mottled appearance. If the joint is large, the solder will look like molten lava that has hardened. The typical cold solder joint is not as smooth as a correctly made joint. Some cold joints even have peaks where solder was applied to the hot iron and was not allowed to flow properly over the joint.

Cold solder joints are repaired in the same fashion as solder bridges: remove the solder from the old joint and try again. If the joint has a sufficient amount of solder, try to fix the problem by applying the iron directly to the joint (not the solder). Wait until the solder melts completely (about 3 to 5 seconds), then remove the iron.

While the solder is cooling, be sure that you don't disturb the work. Movement of the joint while it is cooling will cause a cold or incomplete soldering job.

Incomplete joint

An incomplete joint lacks sufficient solder to bond the component or wire to the circuit board. An incomplete joint can also be caused by dirt, grease, or oil preventing the solder from sticking.

If the joint lacks sufficient solder for the job, reheat it and apply additional solder. Don't add too much or you'll end up with a glob of solder that could cause a short. If the joint is dirty, remove all traces of the solder and clean the joint with an electrical contact cleaner. Use a brush for a thorough cleaning. Let the joint dry and then resolder.

Broken Traces and Leads

The copper foil traces on a circuit board are the wires connecting the components together. Break a trace and that wire is severed, rendering the circuit inoperative. The traces on a circuit board are susceptible to damage in a number of ways.

- The trace could become detached from the circuit board. This can occur if the board is flexed, twisted or exposed to high heat (such as prolonged contact with a soldering iron). Once detached, the trace is brittle and easily broken.
- The thinner the strip of copper foil, the more susceptible it is to breakage. The breakage is more likely to happen when soldering or handling the board, but it can also occur from normal use.

One way to fix a broken trace is to simply apply a solder bridge across the exposed copper. This method applies only if the copper foil has not lifted from the board. If delamination has occurred, the foil should be removed and replaced. You can replace the foil by using a length of bare wire to bridge across the broken section (see Fig. 16-2), or use self-adhesive copper foil strips such as those available from Bishop Graphics.

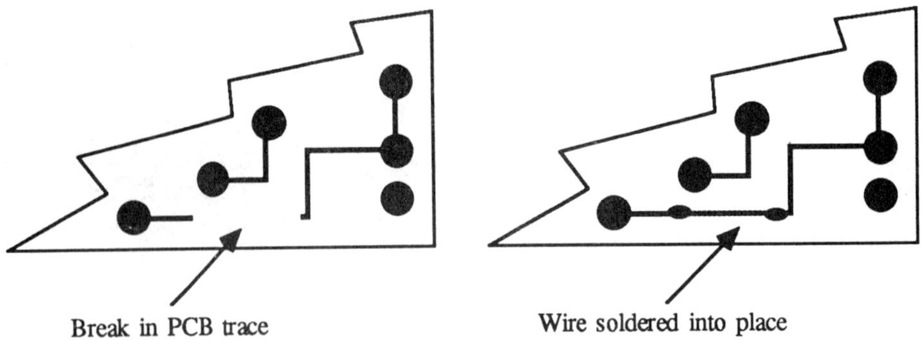

Fig. 16-2. Use a length of wire as a bridge to fix a broken trace.

When soldering on a new wire, use care to avoid disturbing the joints of nearby connections. Before soldering the wire in place, tin the ends and the contact points on the circuit board. Tinning is the action of applying a *light* coat of solder to the metal, which enhances solder flow.

While soldering, attach a heatsink to the wire to help draw heat out of it. This will help avoid melting the first joint while you are soldering the second.

If the trace is long, use a length of insulated wire. You can use wire-wrap wire for most digital circuits; those circuits that consume higher sums of current or those that work under high voltage require a heavier wire.

You might also need to bridge traces if you've made your own printed circuit board and the etchant has oozed under the resist and undercut the copper. The trace could be totally obliterated, or the etching action might dissolve all but a thin sliver of copper foil. In any case, you should repair the trace using one of the methods just outlined.

Replacing Components

Components that go bad must be replaced. This entails:

- Unsoldering the old component.
- Cleaning the joint.
- Inserting the new component.
- Soldering the new component in place.

Once you have identified the component you need to replace, use a desoldering tool to remove the excess solder. You can also wick up the old solder with a copper braid, but this method can become time consuming if you have many components to replace.

You might not be able to remove all of the old solder from the joint. A dab of solder could remain and prevent you from freely removing the component. You can usually free the component leads by heating them up and then pulling or pushing the component out of the hole before the solder cools.

After the component is removed from the circuit board, use the desoldering tool to pick up any remaining solder. When you are ready to insert the new component, wet a cotton ball with isopropyl alcohol and clean the component leads and joint. Wait a few moments for the alcohol to evaporate, and solder the new component into place.

Integrated circuits pose a special desoldering problem. While it's a fairly easy task to remove the bulk of the solder to the leads of the integrated circuit, some solder could remain and prevent you from lifting the IC out of the circuit board. You need to remove as much solder as you can while using care to avoid applying excessive heat to the adjacent components. When you have removed as much of the solder as you possibly can, remove the chip starting from one side. Use a screwdriver to gently pry up on the IC while you heat up the leads from underneath. As the solder melts, you'll be able to lift up one end of the IC and work your way to the other end.

Because of the difficulty in removing ICs, it's always better to use IC sockets on your printed circuit boards. The socket allows a quick and relatively painless method of exchanging ICs without resorting to desoldering and resoldering. Although sockets add a little to the cost of the finished circuit, they are worth the time and energy you'll save if you have to remove an integrated circuit.

17

Using electronic formulas

MOST ELECTRONIC CIRCUITS ARE DESIGNED FIRST ON PAPER AND THEN TESTED using actual components. The design phase requires using mathematical formulas to select the proper values of components. These same formulas are also used when modifying existing circuitry. For example, you can apply the basic Ohm's law formula for direct current to select a resistor so that a light-emitting diode will shine brighter. You would use the formula to provide the greatest amount of current to the LED while avoiding overdriving and destroying it.

This chapter summarizes many of the more common electronic formulas you will encounter in your electronics work. Many of these formulas have been in use for a long time; they are compiled here to provide a convenient reference.

Topic Index

Ohm's Law Formulas

Ohm's law calculates the relationship between power, voltage, current, and resistance.

Unknown Value	Formula
Voltage	$E = IR$
Current	$I = E/R$
Power	$P = EI$
Resistance	$R = E/I$

where E = voltage (in volts)
I = current (in amps)
P = power (in watts)
R = resistance (in ohms)

Example: To find the power in a circuit consuming 100 volts at 10 amps, multiply volts times amps ($100 \times 10 = 1000$). Answer: 1,000 watts.

Figure 17-1 shows a handy reference chart you can use when using ohm's law formulas.

Resistance

The resistance of a single resistor in a circuit is easy to surmise. But resistance changes when you add resistors in parallel or in series.

To calculate two resistors in parallel:

$$R_t = \frac{R_1 \times R_2}{R_1 + R_2}$$

R_1 and R_2 are the values of the two resistors; R_t is the total resistance.

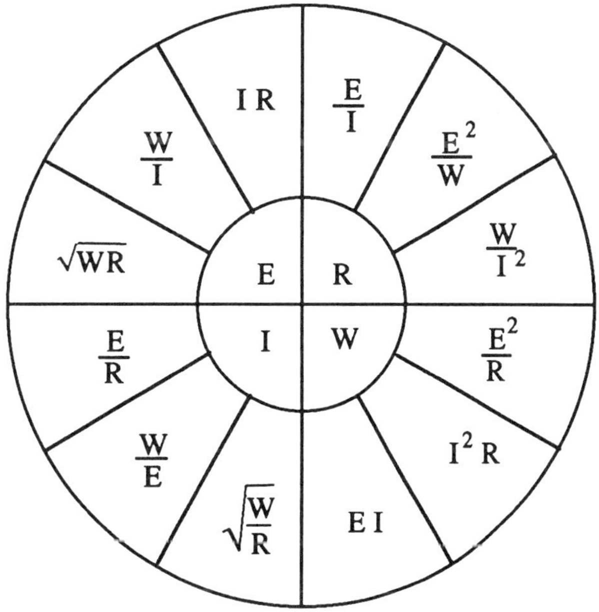

Fig. 17-1. Ohm's law wheel.

To calculate three or more resistors in parallel:

$$R_t = \frac{1}{\frac{1}{R_1} + \frac{1}{R_2} + \frac{1}{R_3} \ldots}$$

R_1, R_2, R_3 (and so forth) are the values of the resistors. R_t is the total resistance.

To calculate resistors in series:

$$R_t = R_1 + R_2 + R_3$$

R_1, R_2, R_3 (and so forth) are the values of the resistors; R_t is the total resistance.

Capacitance

These formulas can be used to calculate total capacitance in a circuit. Note that the formulas are basically the inverse of those for resistors.

To calculate capacitors in parallel:

$$C_t = C_1 + C_2 + C_3 \ldots$$

C_1, C_2, C_3 (and so forth) are the values of the capacitors; C_t is the total capacitance.

To calculate two capacitors in series:

$$C_t = \frac{C_1 \times C_2}{C_1 + C_2}$$

C_1 and C_2 are the values of the two capacitors; C_t is the total capacitance.

To calculate three or more capacitors in series:

$$C_t = \frac{1}{\frac{1}{C_1} + \frac{1}{C_2} + \frac{1}{C_3} \ldots}$$

C_1, C_2, C_3 (and so forth) are the values of the capacitors. C_t is the total capacitance.

To calculate *charge* stored in a capacitor:

$$Q = CE$$

where Q = charge (in coulombs)
C = capacitance (in farads)
E = voltage across capacitor (in volts)

To calculate *energy* stored in a capacitor:

$$W = \frac{CE^2}{2}$$

where W = energy (in joules)
C = capacitance (in farads)
E = voltage (in volts)

Inductance

These formulas can be used to calculate total inductance in a circuit.

To calculate two or more inductors in series (no mutual inductance):

$$L_t = L_1 + L_2 + L_3 \ldots$$

L_1, L_2, L_3 (and so forth) are the values of the inductors; L_t is the total inductance.

To calculate two inductors in parallel (no mutual inductance):

$$L_t = \frac{L_1 \times L_2}{L_1 + L_2}$$

L_1 and L_2 are the values of the two inductors; L_t is the total inductance.

To calculate three or more inductors in series:

$$L_t = \frac{1}{\frac{1}{L_1} + \frac{1}{L_2} + \frac{1}{L_3} \ldots}$$

L_1, L_2, L_3 (and so forth) are the values of the inductors. L_t is the total inductance.

To calculate mutual inductance (when coils of two inductors interact):

$$M = \frac{L_a - L_b}{4}$$

where M = mutual inductance of L_a and L_b (in henrys)
L_a = total inductance of both coils with fields aiding (in henrys)
L_b = total inductance of both coils with fields opposing (in henrys)

To calculate coupled inductance:

In parallel, fields aiding:

$$L_t = \frac{1}{\frac{1}{L_1 + M} + \frac{1}{L_2 + M}}$$

In parallel, fields opposing:

$$L_t = \frac{1}{\frac{1}{L_1 - M} + \frac{1}{L_2 - M}}$$

In series, fields aiding: $L_t = L_1 + L_2 + 2M$

In series, fields opposing: $L_t = L_1 + L_2 - 2M$

where L_t = total inductance (in henrys)
L_1 = value of first inductor (in henrys)
L_2 = value of second inductor (in henrys)

Resonance

Use these formulas to solve for resonant frequency.

To calculate resonant frequency from known coil and capacitor values:

$$f_r = \frac{1}{2\pi\sqrt{LC}}$$

where f_r = resonant frequency (in hertz)
L = inductance (in henrys)
C = capacitance (in farads)

To calculate value of inductor when resonant frequency is known:

$$L = \frac{1}{4\pi^2 f_r^2 C}$$

where L = inductance (in henrys)
f_r = resonant frequency (in hertz)
C = capacitance (in farads)

Conductance

Use this formula to calculate the ability of a component to conduct electricity (dc circuits only):

$$G = \frac{1}{R}$$

where G = conductance (in siemens)
R = resistance (in ohms)

Reactance

Reactance is the opposition to the flow of alternating current within a circuit. There are two general forms of reactance: capacitive and inductive.

To calculate capacitive reactance:

$$R_c = \frac{1}{2\ \pi\ fC}$$

where R_c = reactance (in ohms)
f = frequency (in hertz)
C = capacitance (in farads)

To calculate inductive reactance:

$$R_c = 2\pi fL$$

where R_c = reactance (in ohms)
f = frequency (in hertz)
L = inductance (in henrys)

Q Factor

The Q factor is the ratio between reactance and resistance.

To calculate Q factor with capacitor and resistor in series:

$$Q = \frac{1}{wRC}$$

To calculate Q factor with inductor and resistor in series:

$$Q = \frac{wL}{R}$$

where Q = Q factor ratio
w = $2\pi f$; f is the frequency (in hertz)
R = resistance (in ohms)
C = capacitance (in farads)

Admittance

Admittance is the ease with which an alternating current flows through a circuit.

To calculate the admittance of a series circuit:

$$Y = \frac{1}{\sqrt{R^2 + X^2}}$$

where Y = admittance (in siemens, the unit of conductance)
R = resistance (in ohms)
X = reactance (in ohms)

The reciprocal of impedance can also be used to calculate admittance, using the formula:

$$Y = \frac{1}{Z}$$

where Y is the admittance (in siemens), and Z is the impedance (in ohms).

Susceptance

Calculate the susceptance of a series circuit with the formula:

$$B = \frac{X}{R^2 + X^2}$$

where B = susceptance (in siemens)
X = reactance (in ohms)
R = resistance (in ohms)

Impedance

Two formulas can be used for calculating impedance in a dc circuit for both parallel and series circuits without regard to the combinations of inductance, capacitance, and resistance.

To calculate impedance in a parallel circuit:

$$Z = \frac{1}{\sqrt{G^2 + B^2}}$$

To calculate impedance in a series circuit:

$$Z = \sqrt{R^2 + X^2}$$

where Z = total impedance (in ohms)
G = total conductance or reciprocal of total parallel resistance (in siemens)
X = total reactance (in ohms)
R = total resistance (in ohms)
B = total susceptance (in siemens)

Units of Energy

The watt-hour is one of the most practical units used to measure energy, which is the ability of a device or circuit to do work. Watt-hours are calculated by multiplying the power of the circuit (in watts) by the length of time the circuit is on:

$$\text{watt-hours} = P \times T$$

where P = power (in watts)
T = time power is dissipated (in hours)

To calculate watt-seconds (or joules), divide watt-hours by 60.

RMS, Average, Peak, and Peak-to-Peak ac Voltage and Current

Rms, average, peak, and peak-to-peak calculations of an ac circuit can be quickly calculated using the multipliers in Table 17-1.

Table 17-1. Converting Voltage/Current Values

	Multiply Factor to Get			
To Find	**RMS**	**Average**	**Peak**	**Peak-to-Peak**
Average	1.11	- - -	1.57	3.14
Rms	- - -	0.9	1.414	2.828
Peak	0.707	0.637	- - -	2.0
Peak-to-peak	0.3535	0.32	0.5	- - -

Time Constants

Time constants are often used in electronic circuits to provide time delays or stretch the timing of signals.

To calculate the time constant for a resistor-capacitor circuit:

$$T = RC$$

where T = time (in seconds)
R = resistance (in ohms)
C = capacitance (in farads)

To calculate the time constant for a resistor-inductor circuit:

$$T = \frac{L}{R}$$

where T = time (in seconds)
L = inductance (in henrys)
R = resistance (in ohms)

Frequency and Wavelength

The frequency of a signal is directly proportional to its wavelength, as expressed in the following formulas.

To calculate frequency of a signal:

$$\text{frequency} = \frac{300{,}000}{\text{wavelength}}$$

To calculate wavelength of a signal:

$$\text{wavelength} = \frac{300{,}000}{\text{frequency}}$$

Wavelength is expressed in meters, and frequency is expressed in kilohertz.

Electronic Conversion Factors

Use Table 17-2 to easily convert between various electronics quantities.

Table 17-2. Conversion Factors

To Convert	Into	Multiply by
ampere-hours	coulombs	3600
coulombs	ampere-hours	2.778×10^{-4}
electron volts	joules	6.24×10^{18}
ergs	joules	10^{-7}
ergs per second	watts	10^{-7}
faradays	ampere-hours	26.8
gammas	teslas	10^{-9}
gausses	teslas	10^{-4}
horsepower	kilowatts	0.746
horsepower	watts	746
horsepower-hours	kilowatt-hours	0.7457
joules	electronvolts	0.624×10^{18}
joules	ergs	10^{7}
joules	kilowatt-hours	2.778×10^{-7}
joules	watt-hours	2.778×10^{-4}
kilowatts	horsepower	1.341
kilowatt-hours	horsepower-hours	1.341
kilowatt-hours	joules	3.6×10
kilowatt-hours	watt-hours	10^{3}
teslas	gammas	10^{9}
teslas	gausses	10^{4}
watts	ergs per second	10^{7}
watts	horsepower	1.341×10^{-3}
watt-hours	joules	3600
watt-hours	kilowatt-hours	10^{-3}

Letter Symbols

A ampere; amp
Ah ampere-hour
c/s cycles per second (obsolete; see Hz)

dB	decibel
eV	electronvolt
F	farad
GHz	gigahertz
H	henry
hp	horsepower
Hz	hertz
J	joule
k	kilo- (thousand)
kHz	kilohertz
kV	kilovolt
kW	kilowatt
kWh	kilowatt hour
M	mega (million)
MeV	megaelectron volt
MΩ	megohm
mA	milliampere
mH	millihenry
ms	millisecond
mV	millivolt
mW	milliwatt
nF	nanofarad
ns	nanosecond
pF	picofarad
S	siemens
V	volt
V/A	voltampere
W	watt
μ	micro- (millionth)
μA	microampere
μF	microfarad
μH	microhenry
μs	microsecond

Abbreviations

ac	alternating current
AM	amplitude modulation
amp	ampere, amplifier
ant	antenna
ASCII	American Standard Code for Information Interchange
assy	assembly
aud	audio

aux	auxiliary
avg	average
AWG	American Wire Gauge
bw	bandwidth
cap	capacitor
cath	cathode
C	capacitance, capacitor, collector, coulomb
dc	direct current
DPDT	double-pole, double-throw (switch)
DPST	double-pole, single-throw (switch)
emf	electromotive force
EMT	electrical metallic tubing
ERP	effective radiated power
E	emitter, voltage
f	frequency
FM	frequency modulation
HF	high frequency
HT	high tension (high voltage and high current)
HV	high voltage
ID	inside diameter
IF	intermediate frequency
in	input
I/O	input/output
IR	infrared
I	current
IC	integrated circuit
K	kelvin, dielectric constant
l	length
L	inductor, inductance
LF	low frequency
LP	low-pass (filter)
LSB	lower sideband
LSI	large-scale integration
mic	microphone
mom	momentary (switch)
NC	no connection, normally closed
neg	negative
neut	neutral
NO	no, normally open
nom	nominal (representative)
norm	normal
osc	oscillator, oscillator, oscilloscope
out	output

PC	printed circuit
PCB	printed circuit board
PCM	pulse-code modulation
PF	power factor
pk	peak
pos	positive
pot	potentiometer
preamp	preamplifier
PS	power supply
PU	pickup
pri	primary
pwr	power
Q	merit of capacitor or coil
rcv	receive
rcvr	receiver
rect	rectifier
ref	reference
res	resistor
RF	radio frequency
RFI	radio frequency interference
RLY	relay
rms	root mean square
rmt	remote
rot	rotate
rpm	revolutions per minute
rps	revolutions per second
rpt	repeat
R	resistance, resistor
RC	resistance/capacitance
RL	resistance/inductance
SB	sideband
sec	second, secondary
sel	selector
sft	shaft
sig	signal
SNR	signal-to-noise ratio
sol	solenoid
SPDT	single-pole, double-throw (switch)
spk	speaker
spkr	speaker
SPST	single-pole, single-throw (switch)
sq	square
stby	standby

subassy	subassembly
sw	switch, shortwave
t, T	time, temperature
tel	telephone
UHF	ultrahigh frequency
UF	ultrasonic frequency
V	voltage
vid	video
VLF	very low frequency
VOM	volt-ohmmeter
VTVM	vacuum-tube voltmeter
wdg	winding
X	reactance
Z	impedance

18

Circuit grab bag

MOST ELECTRONIC CIRCUITS ARE COMPOSED OF COMMON BUILDING BLOCKS: A 555 timer here, a filter there, an op amp working as a signal preamplifier. These common building blocks can be used by themselves—a simple 555 astable oscillator used as a 1 kHz tone source, for example—or they can be a part of a larger and more complex circuit such as a lightwave communications system, a radar detector, or a television signal descrambler.

This chapter presents numerous circuits and circuit fragments you can use in your electronic tinkering. You'll find circuits for timers, oscillators, multivibrators, op amps, filters, LED, lamp, and relay drivers, counters, switch debouncers, and more.

Circuit Index

Timers

Astable

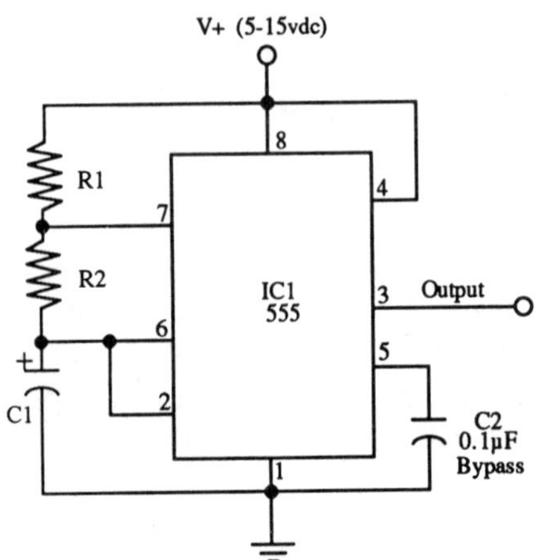

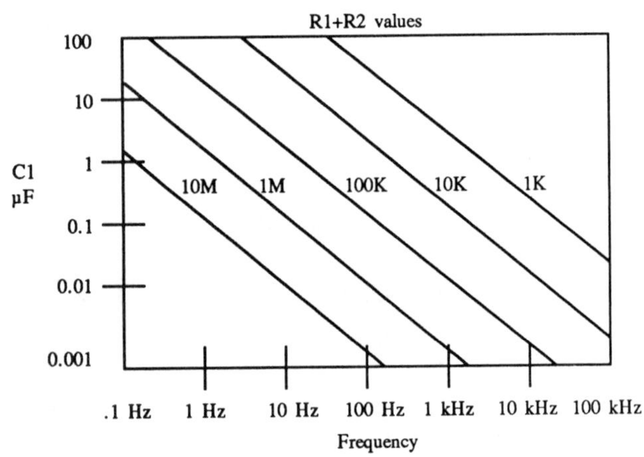

Monostable

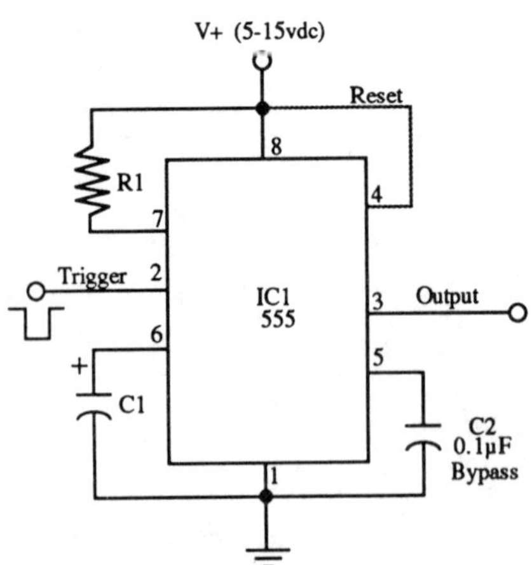

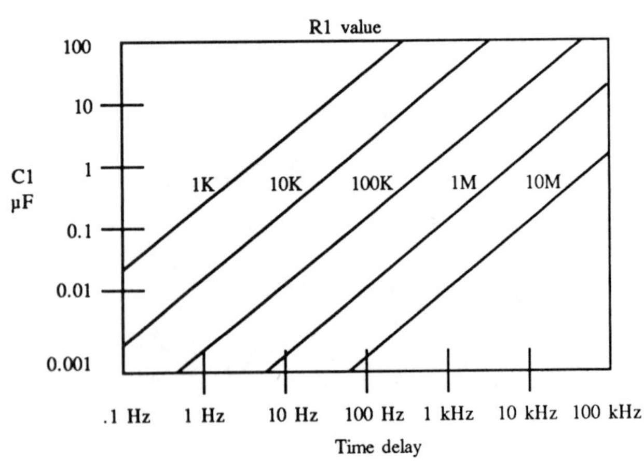

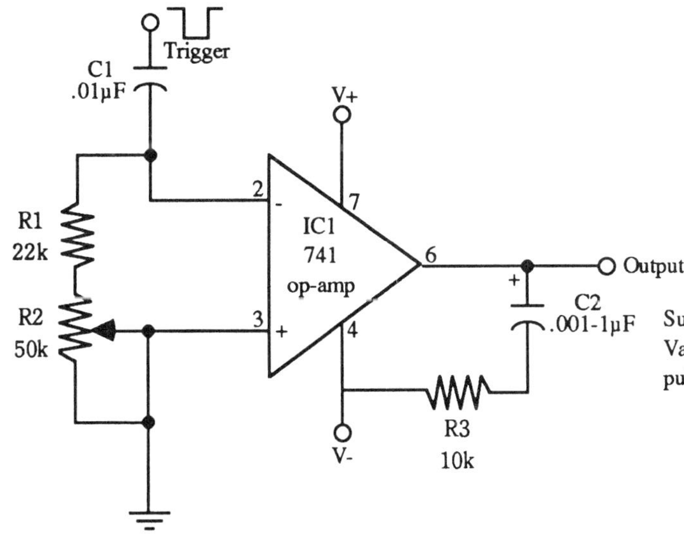

Supply voltage to ± 9vdc. Vary R2 and C2 for different pulse duration

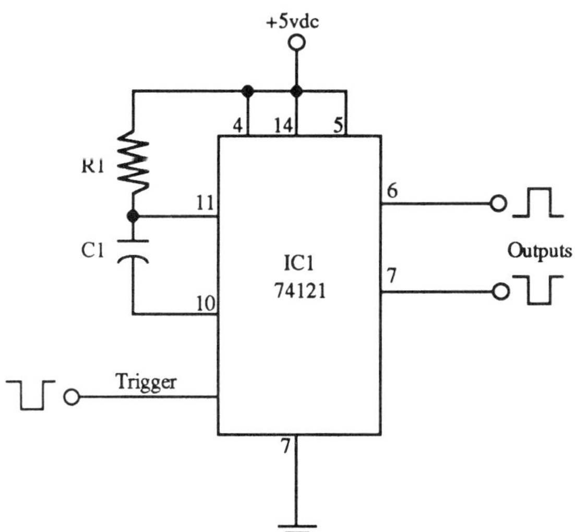

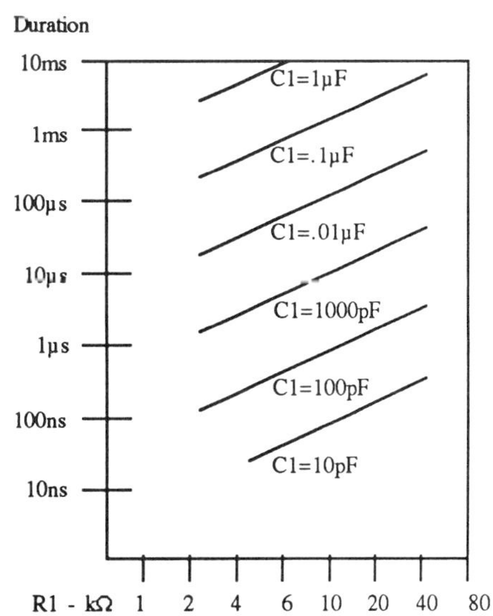

Oscillators

Op amp

Square wave oscillator

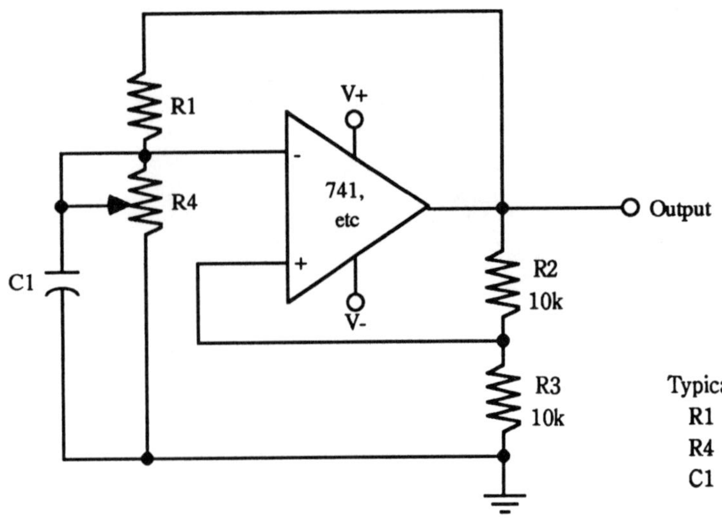

Typical component values:

R1 - 10k
R4 - 50k
C1 - 0.001 to 1 μ

Low-frequency square wave

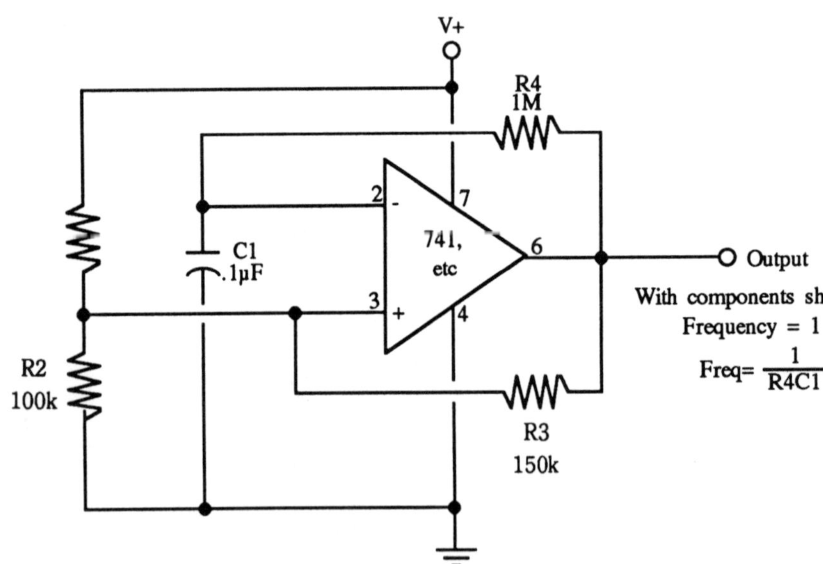

With components shown:
Frequency = 1 Hz

$$\text{Freq}=\frac{1}{R4C1}$$

Triangle wave generator

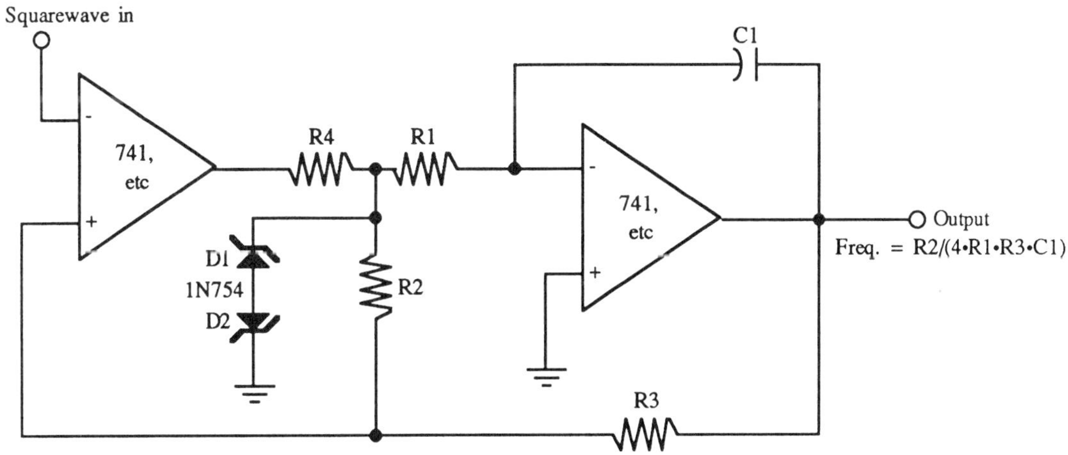

Sine/Cosine sinewave oscillator

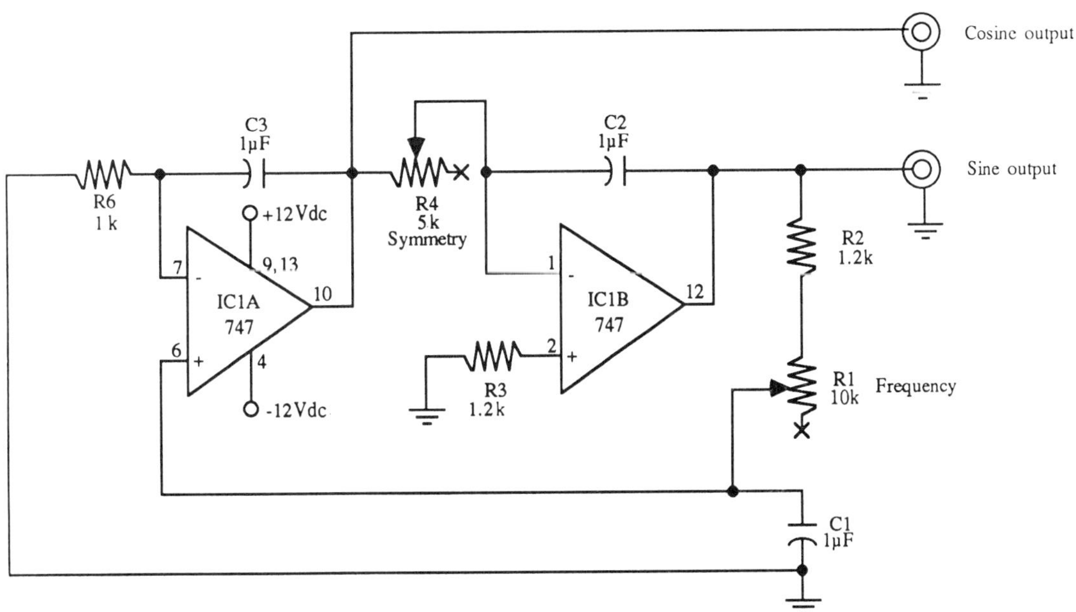

Sinewave oscillator; adjustable

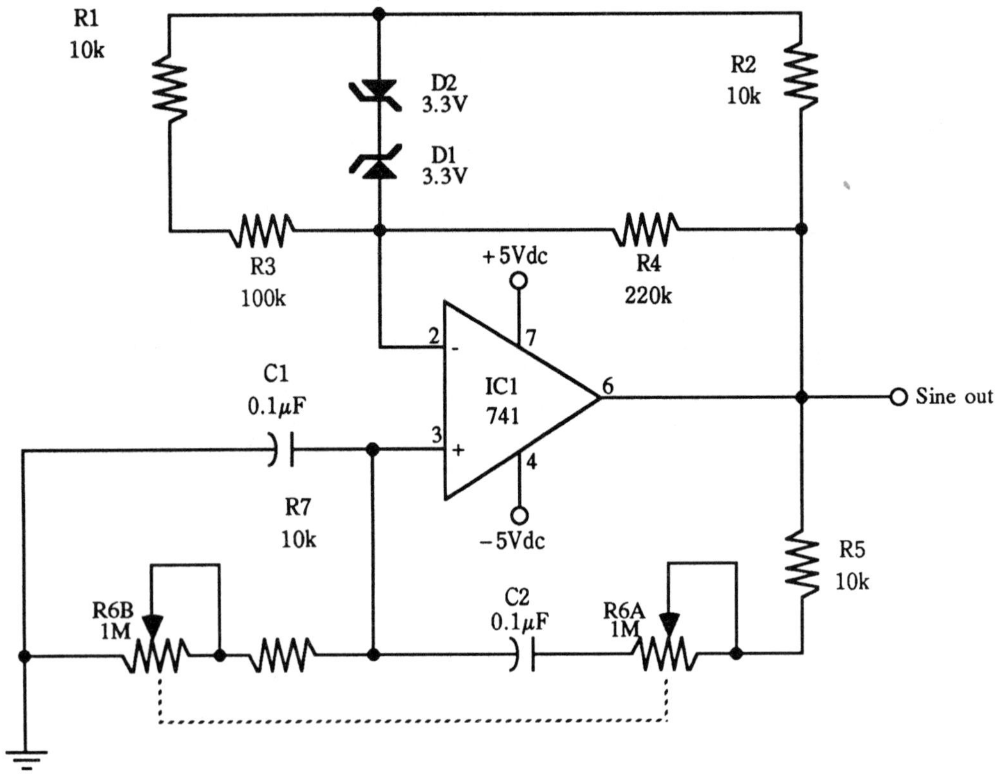

Logic

Logic IC astable oscillator (square wave)

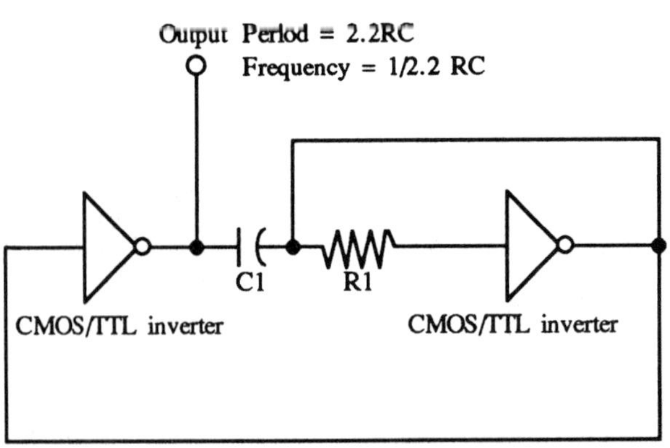

Variable-frequency square-wave oscillator

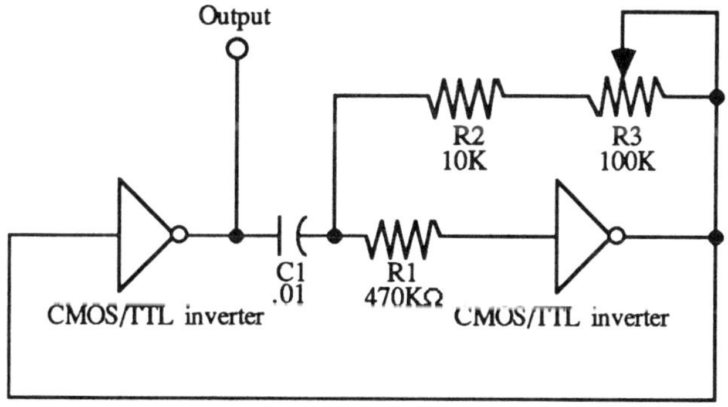

Oscillator with symmetrical square-wave output

Period = 2.2RC
Frequency = 1/2.2 RC

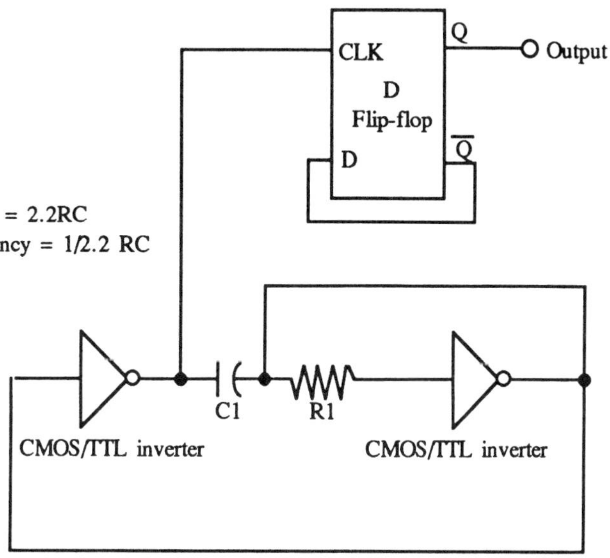

Gated oscillator

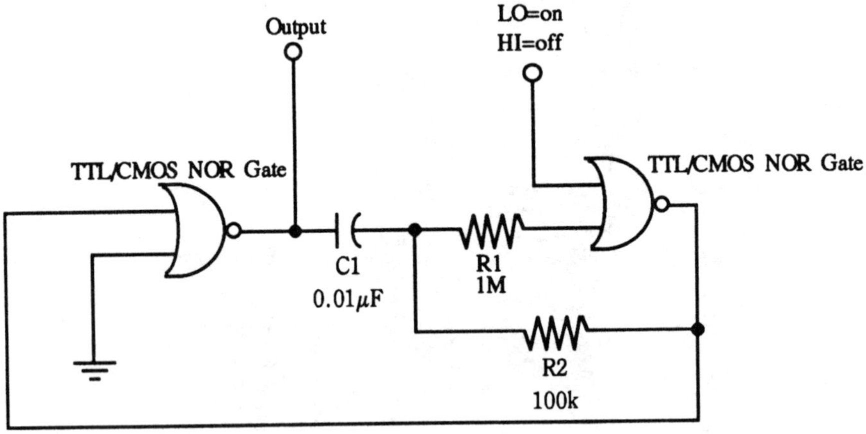

Gated oscillator

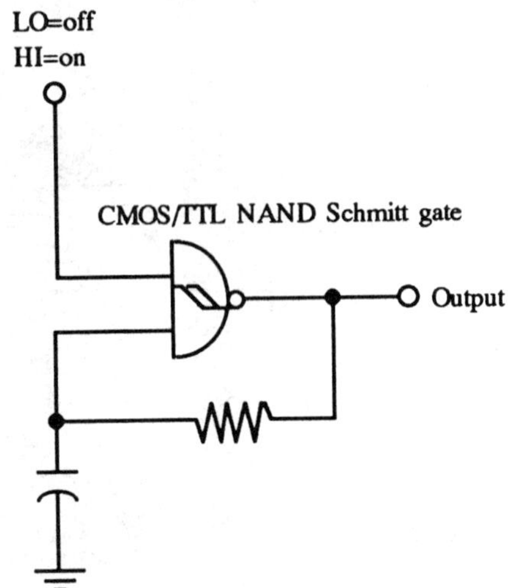

Oscillator using Exar XR-2206 IC

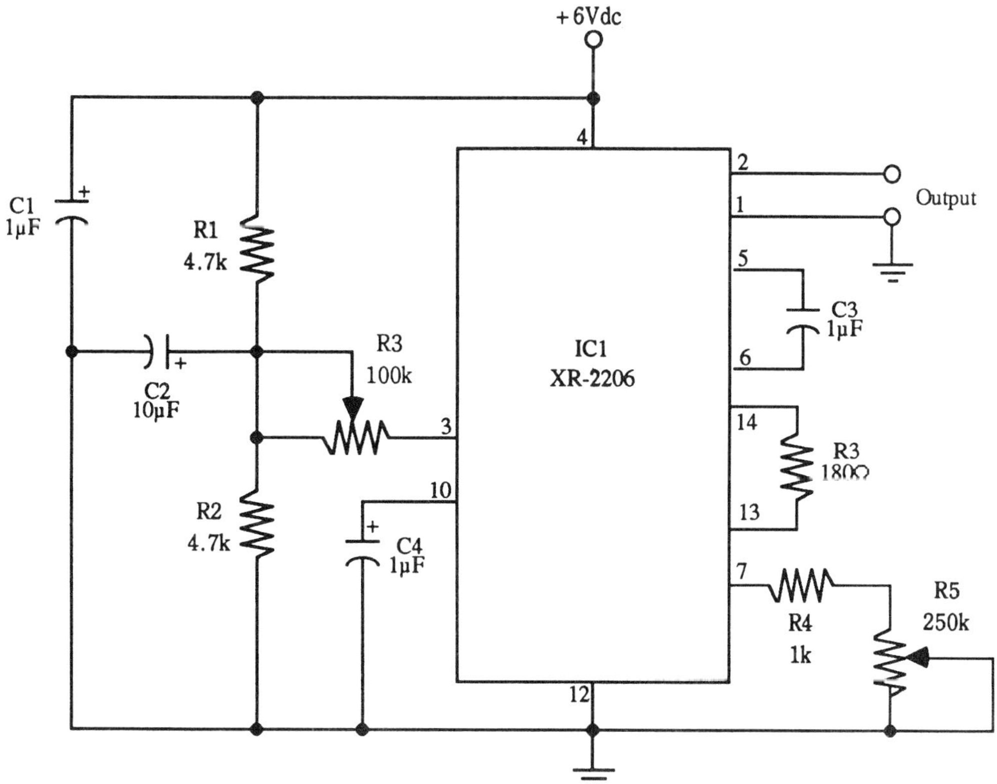

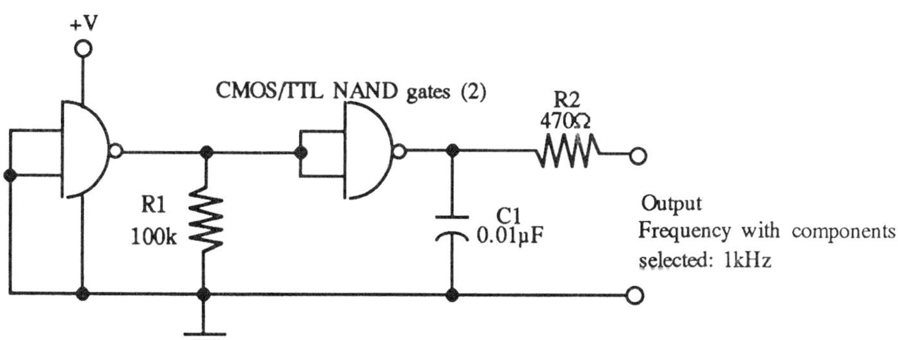

Filters

Passive filter; low-pass

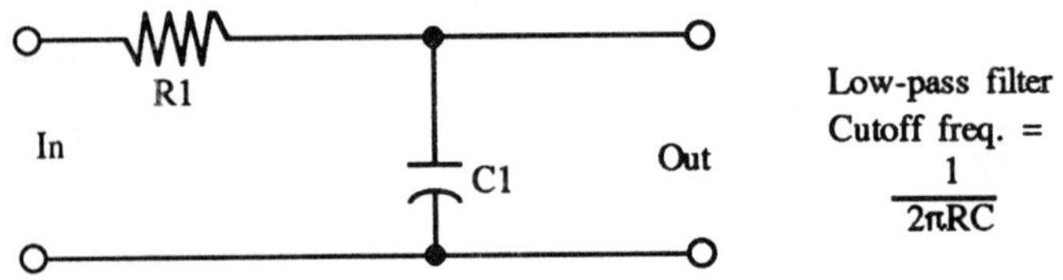

Passive filter; high-pass

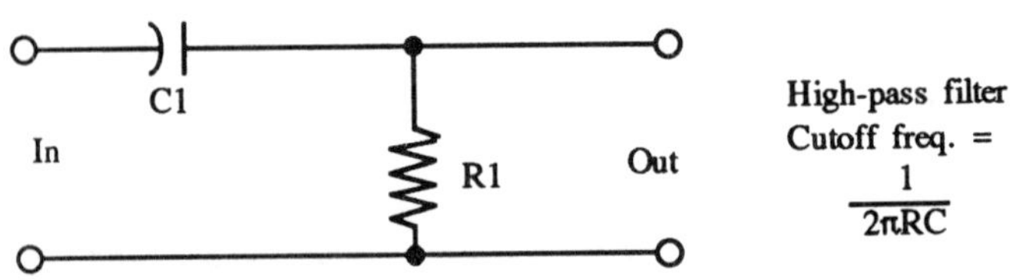

10kHz active filter

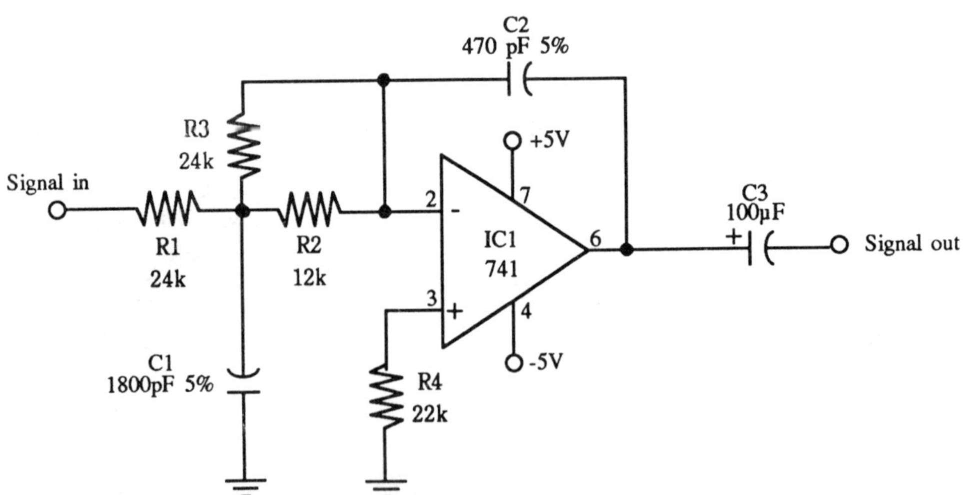

Active low-pass filter

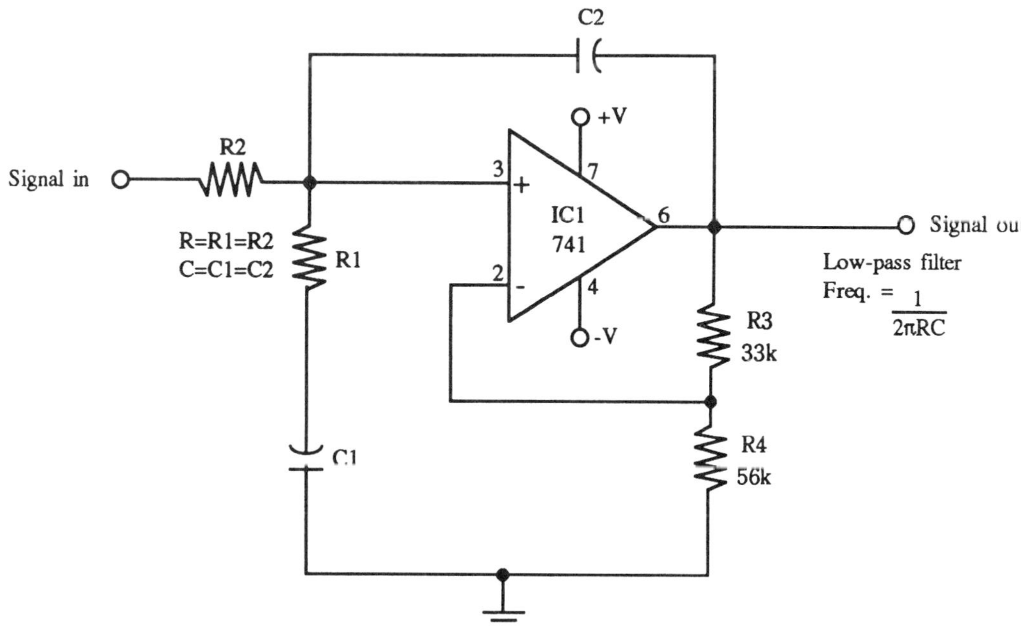

Active high-pass filter

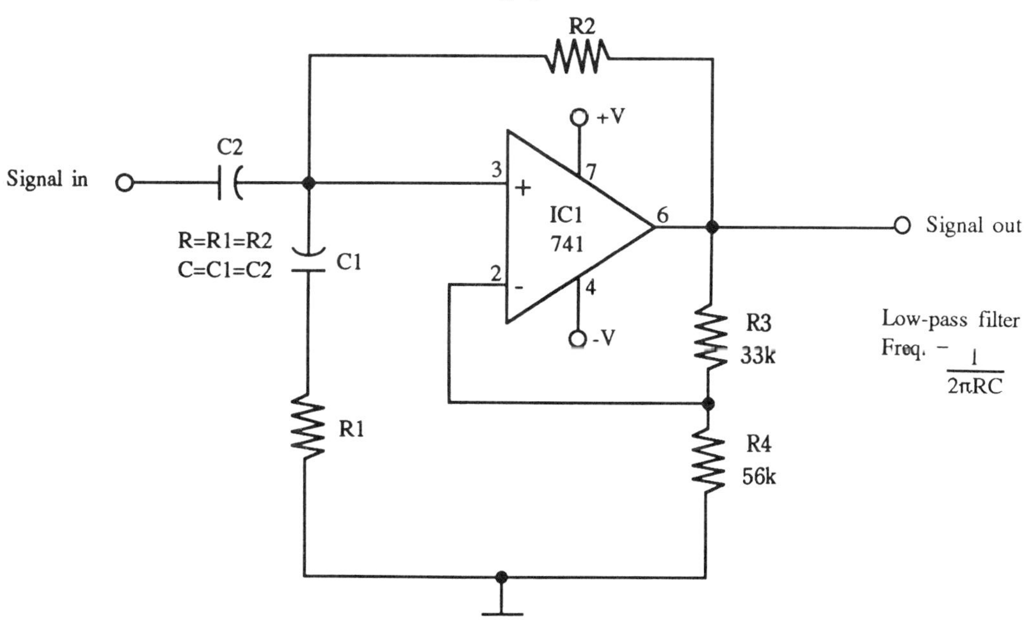

Crystal Timebase

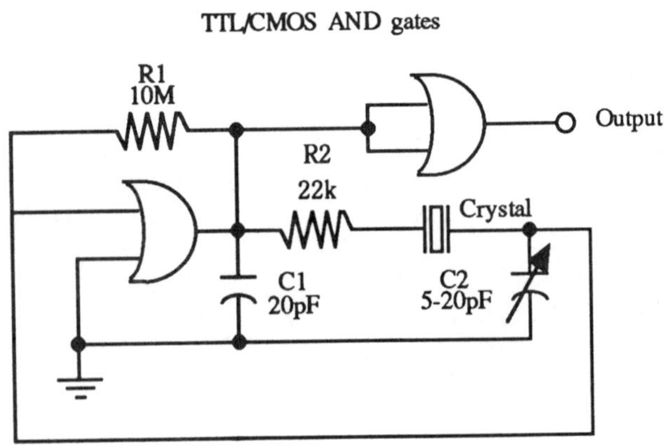

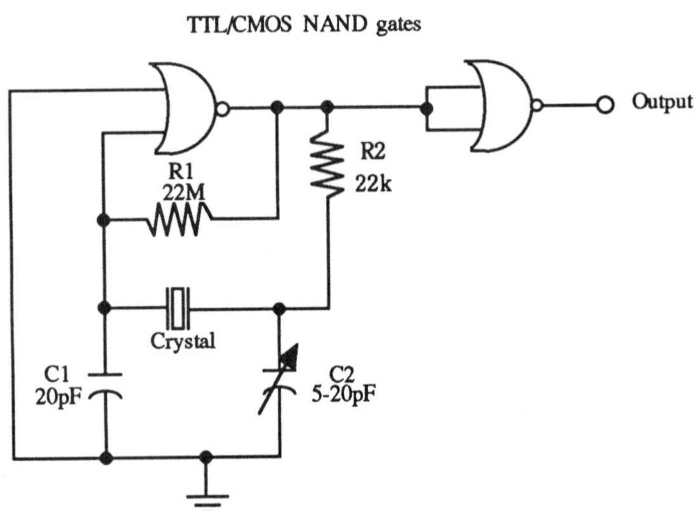

Counters

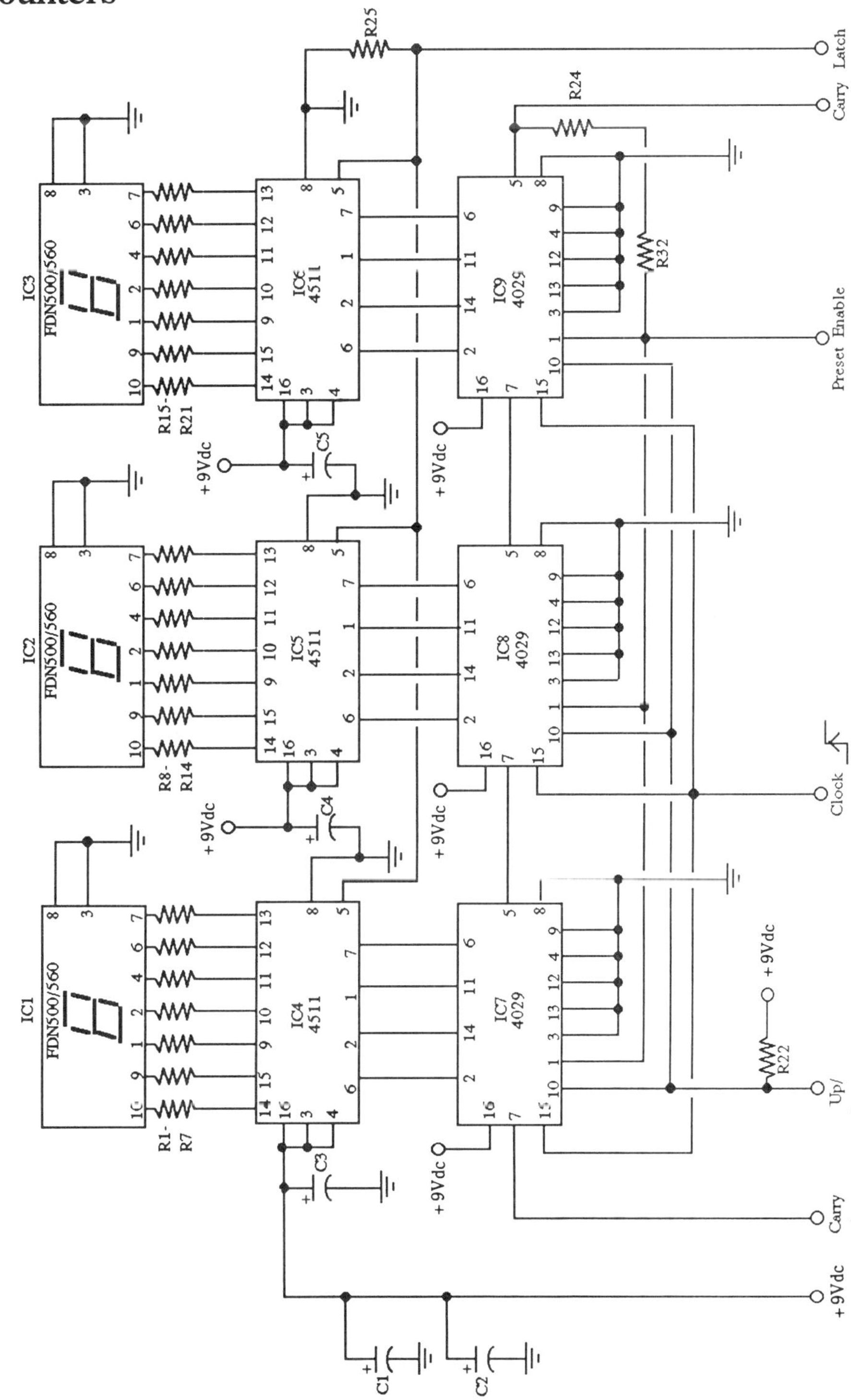
IC1
FDN500/560
IC2
FDN500/560
IC3
FDN500/560
IC4
4511
IC5
4511
IC6
4511
IC7
4029
IC8
4029
IC9
4029
R1-
R7
R8-
R14
R15-
R21
R22
R24
R25
R32
C1
C2
C3
C4
C5
+9Vdc
Latch
Enable
Carry
In
Preset Enable
Clock
Up/
Down
Carry
Out

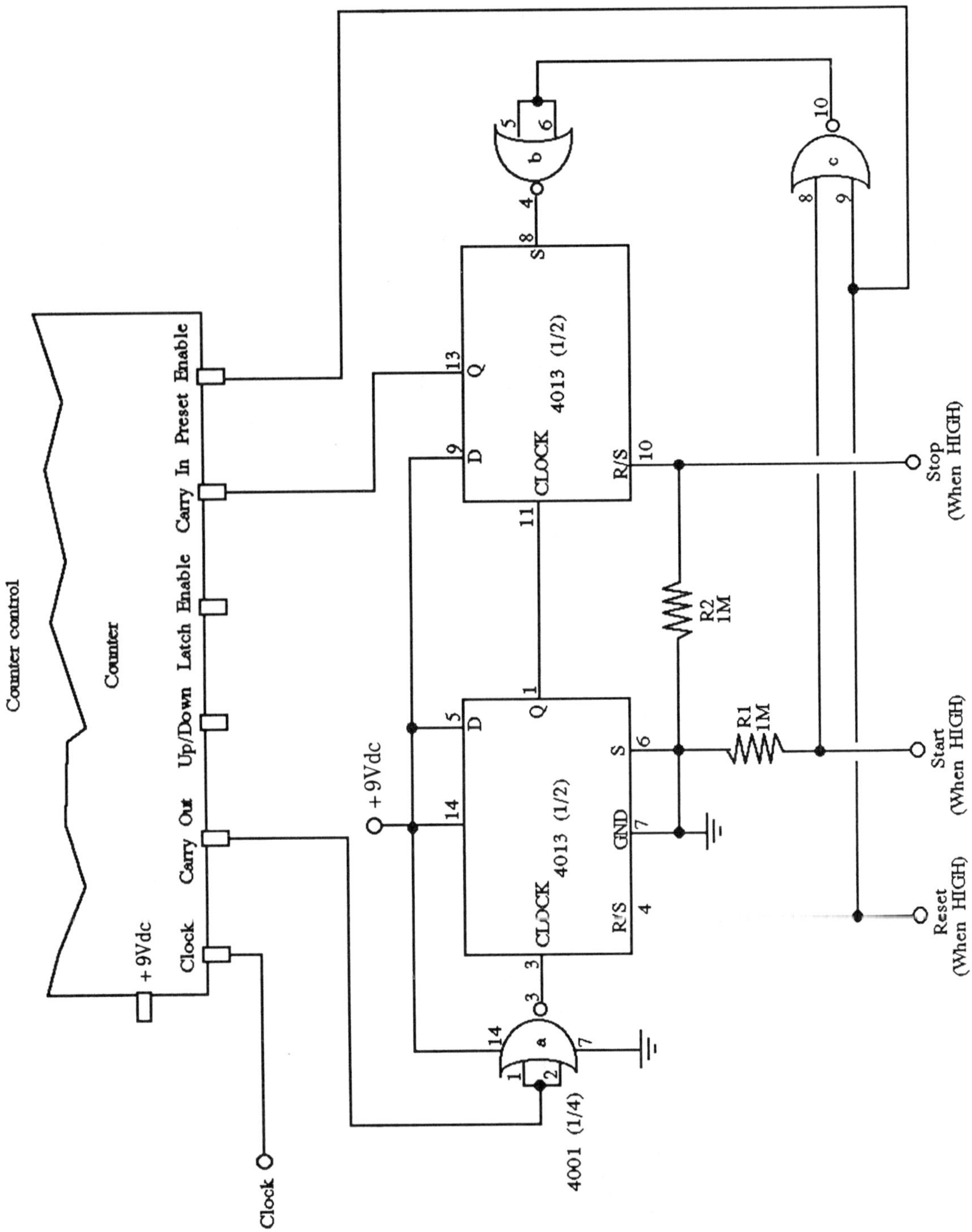
Counter control
Counter
+9Vdc
Clock
Carry Out
Up/Down
Latch Enable
Carry In
Preset Enable
Clock
4001 (1/4)
+9Vdc
4013 (1/2)
4013 (1/2)
R1
1M
R2
1M
Reset
(When HIGH)
Start
(When HIGH)
Stop
(When HIGH)

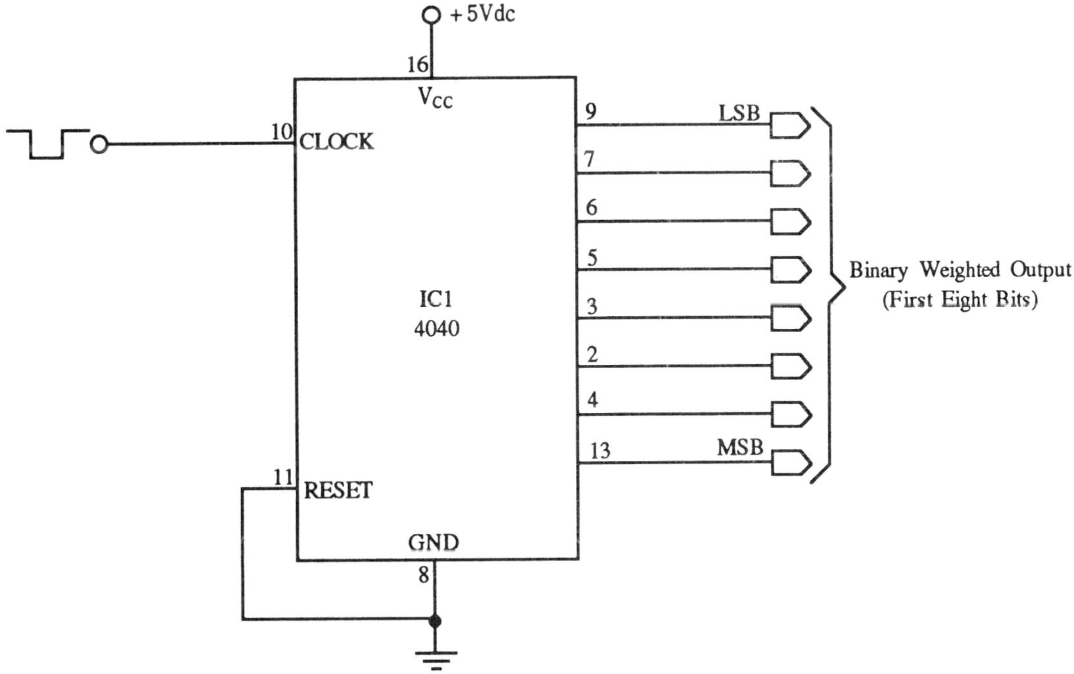
+5Vdc
16
Vcc
10
CLOCK
9
LSB
7
6
5
3
2
4
13
MSB
IC1
4040
11
RESET
GND
8
Binary Weighted Output
(First Eight Bits)

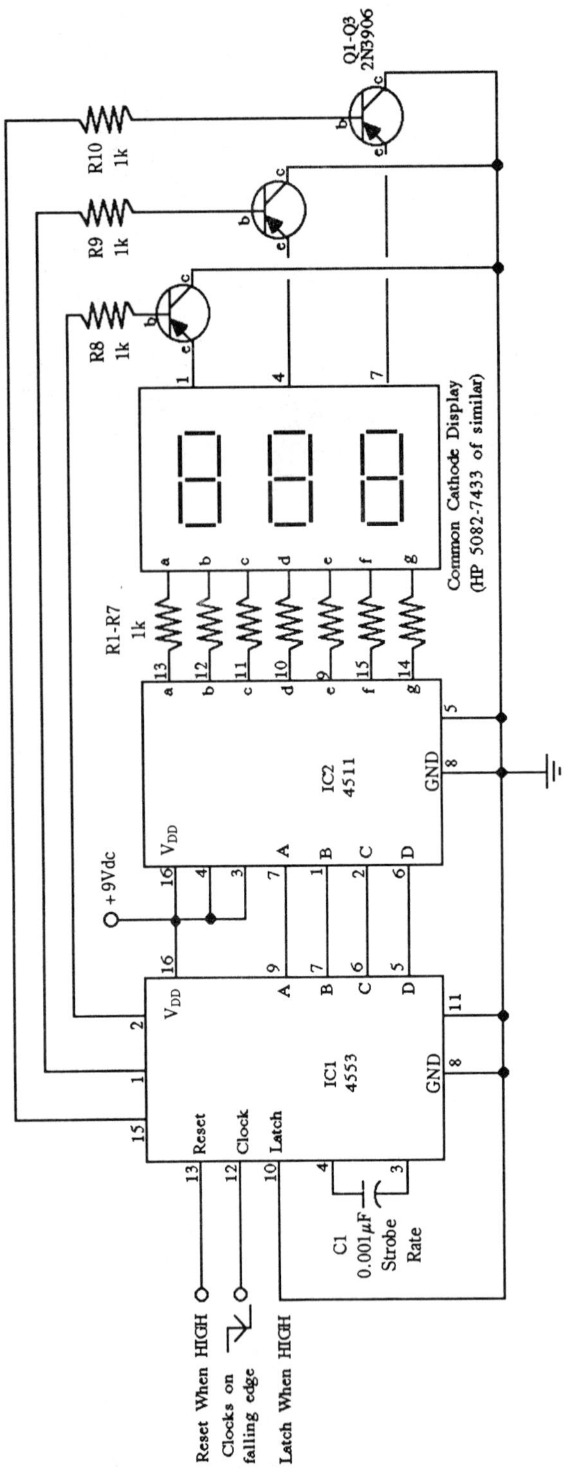
Q1-Q3
2N3906
R10
1k
R9
1k
R8
1k
R1-R7
1k
Common Cathode Display
(HP 5082-7433 of similar)
IC2
4511
IC1
4553
+9Vdc
C1
0.001µF
Strobe
Rate
Reset When HIGH
Clocks on
falling edge
Latch When HIGH

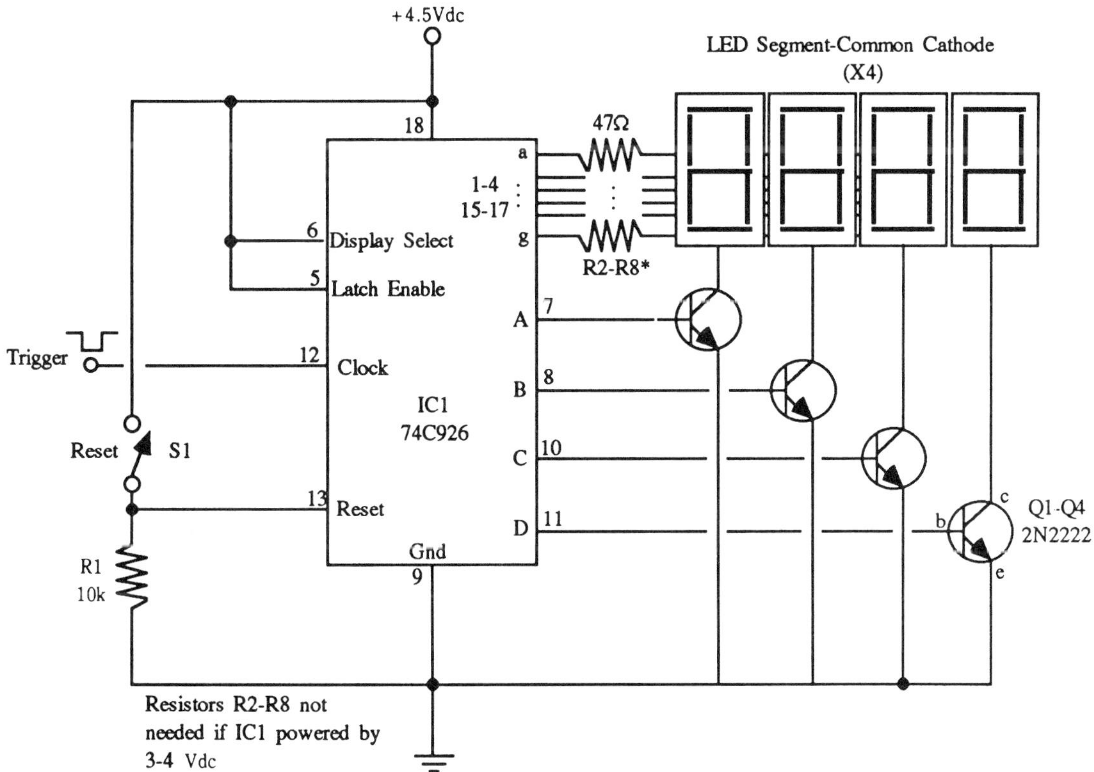
+4.5Vdc
LED Segment-Common Cathode
(X4)
47Ω
18
a
1-4
15-17
g
R2-R8*
6
Display Select
5
Latch Enable
A
7
Trigger
12
Clock
B
8
IC1
74C926
Reset
S1
C
10
13
Reset
D
11
c
Q1-Q4
2N2222
b
e
Gnd
9
R1
10k
Resistors R2-R8 not
needed if IC1 powered by
3-4 Vdc

Tone Decoder

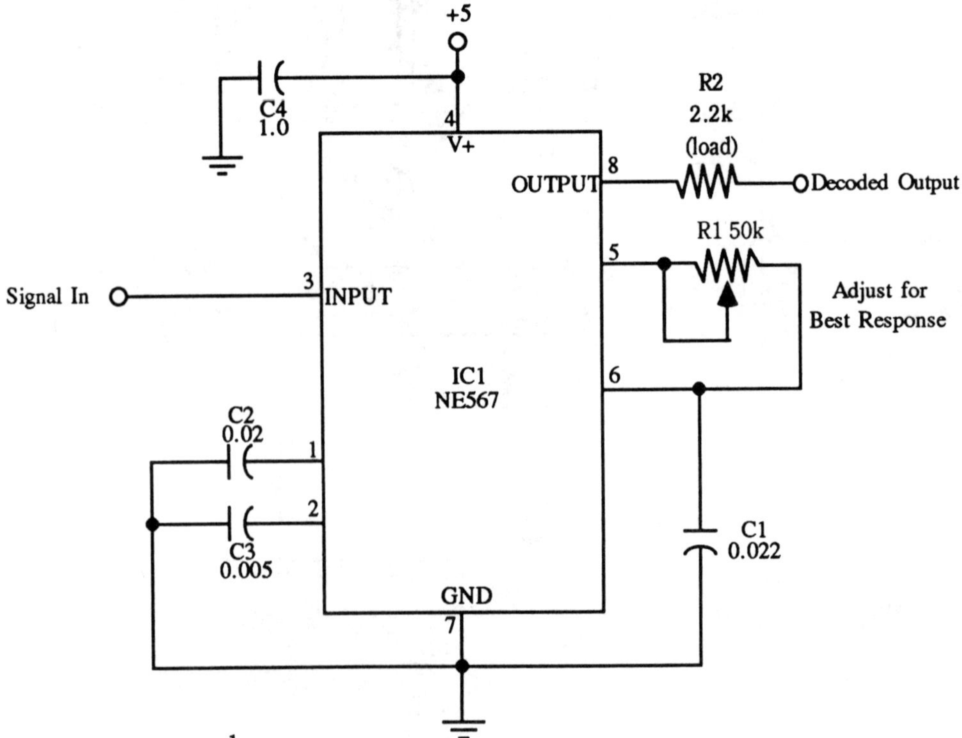

$$f_o = \frac{1}{1.1R1C1}$$

$$\sqrt{\text{Bandwidth} = 1070\frac{V_{in}}{f_oC2}} \text{ in \% of } f_o$$

Components selected for approx. 1kHz response

Dividers

Divide-by-2

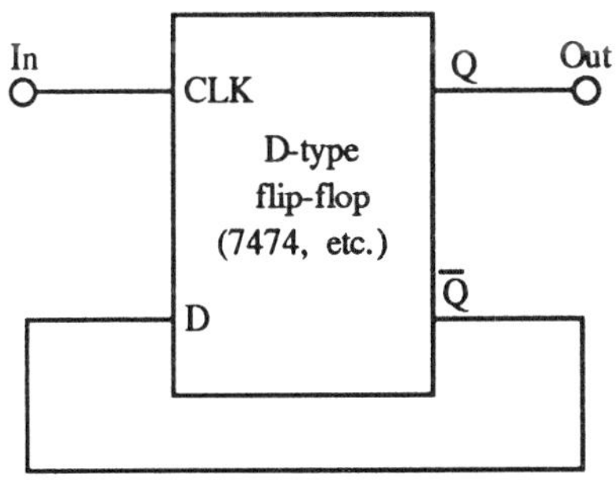

Divide-by-5

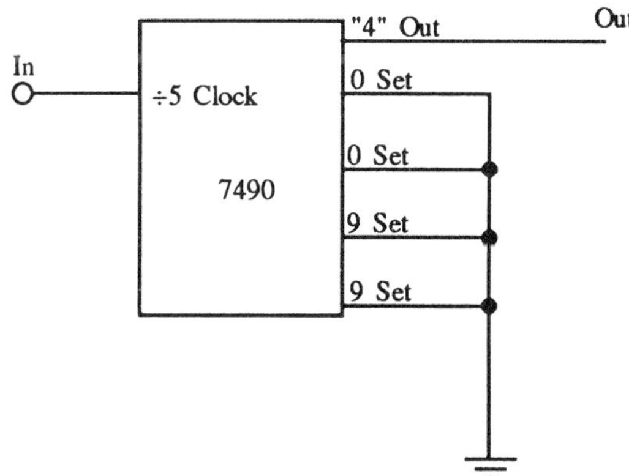

Divide-by-4

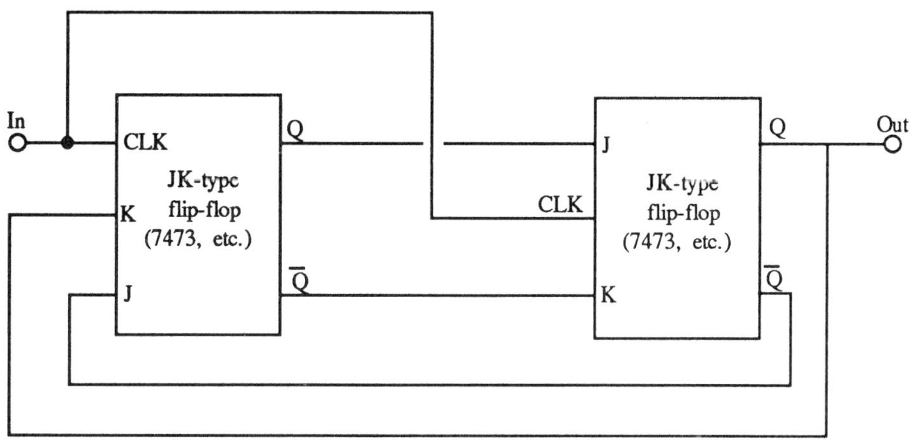

Divide-by-6

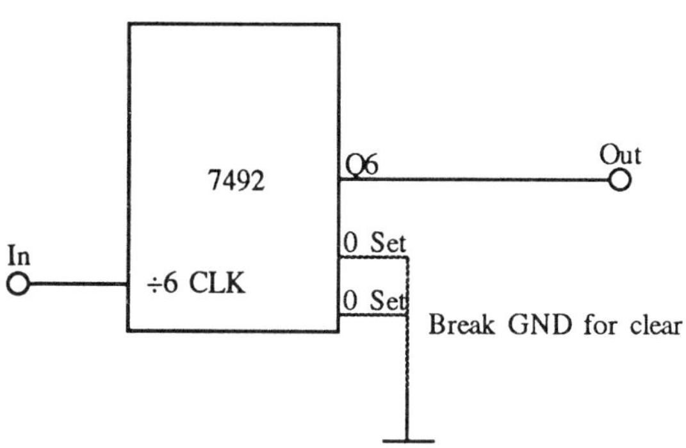

Divide-by-10

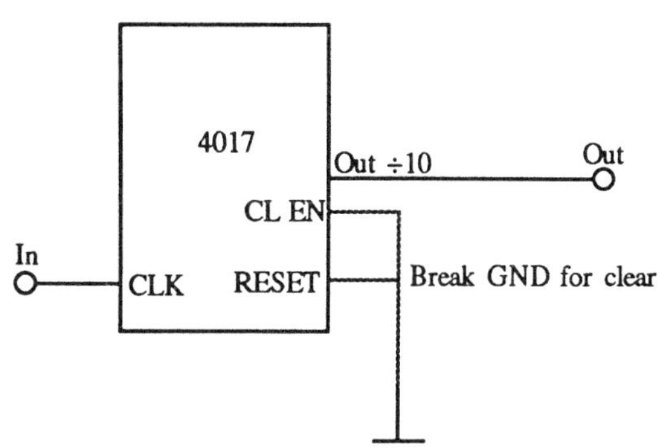

Amplifiers

Op amp

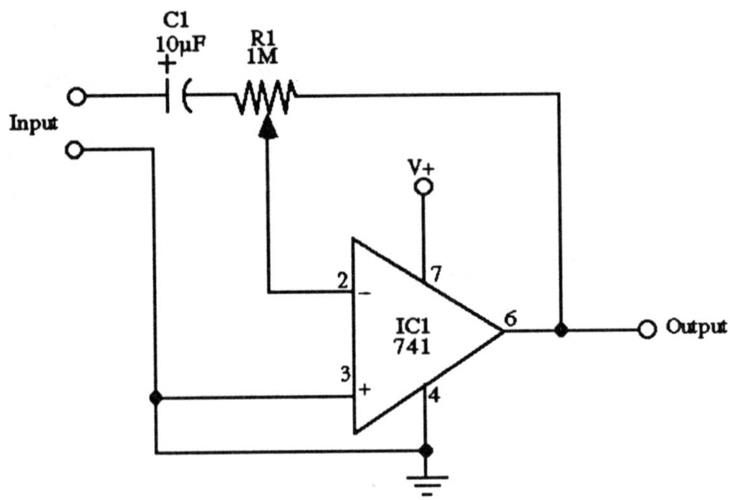

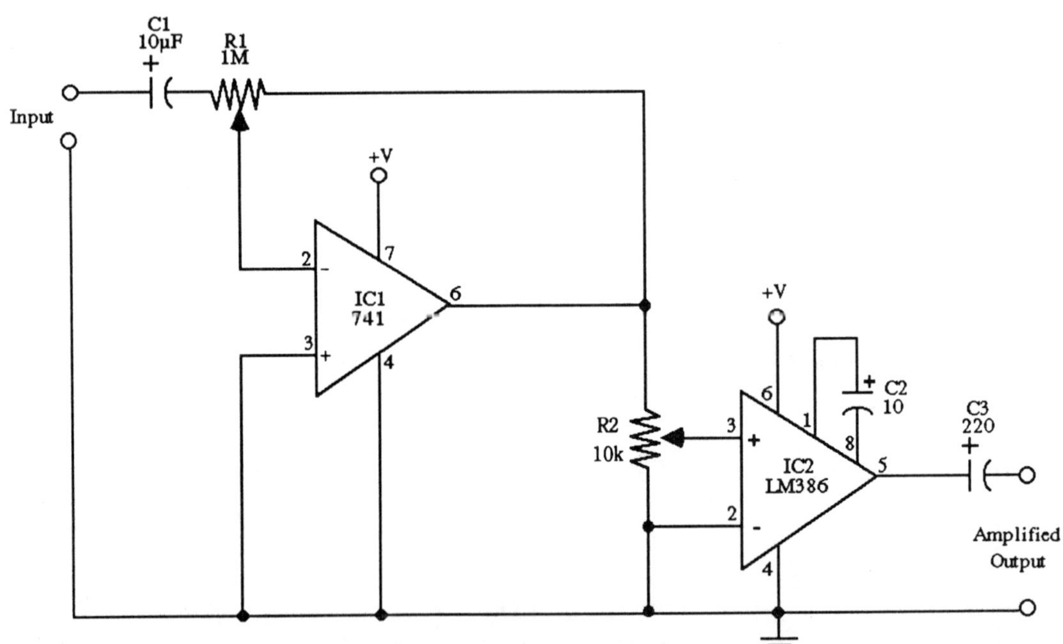

Transistor

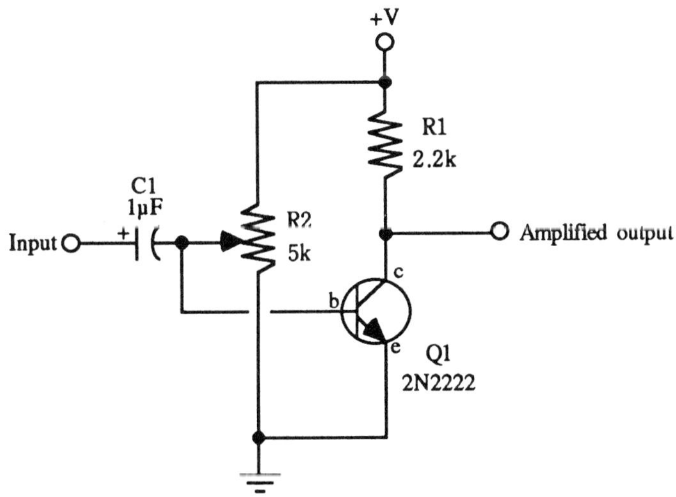

Specialty

LM380 mini-power audio amplifier

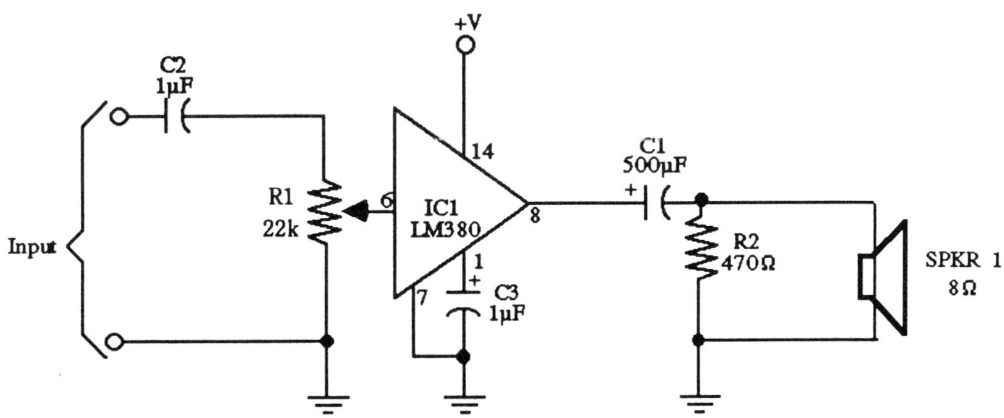

Audio amplifier; gain of 50

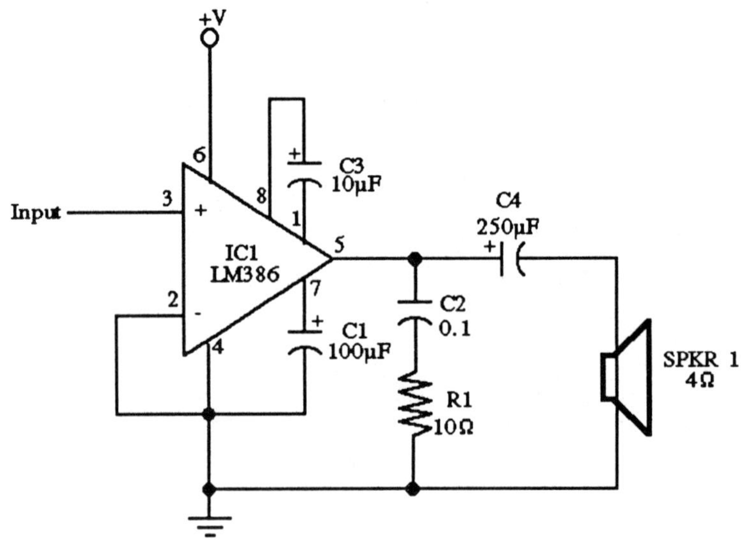

Audio amplifier; gain of 50

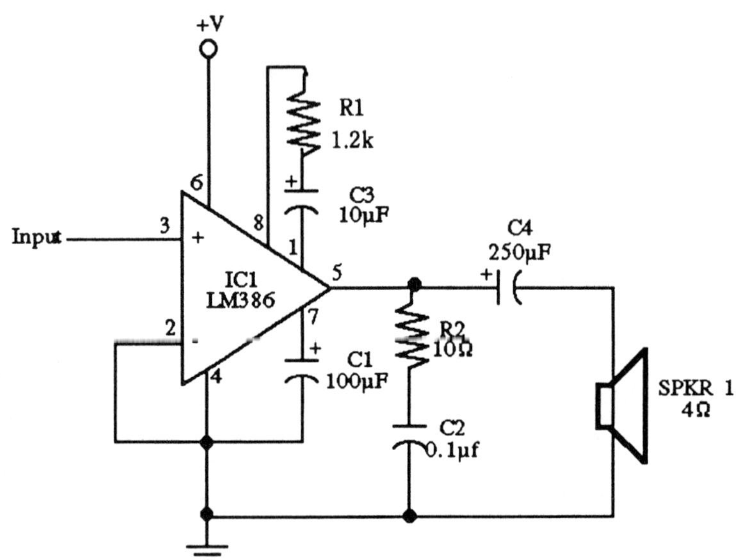

8-watt audio amplifier

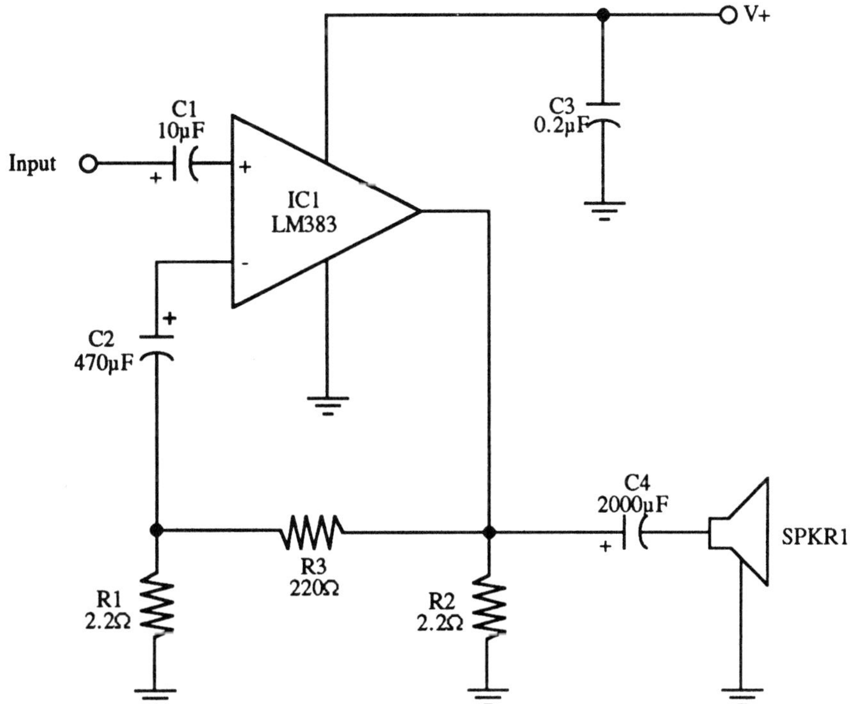

16-watt audio amplifier

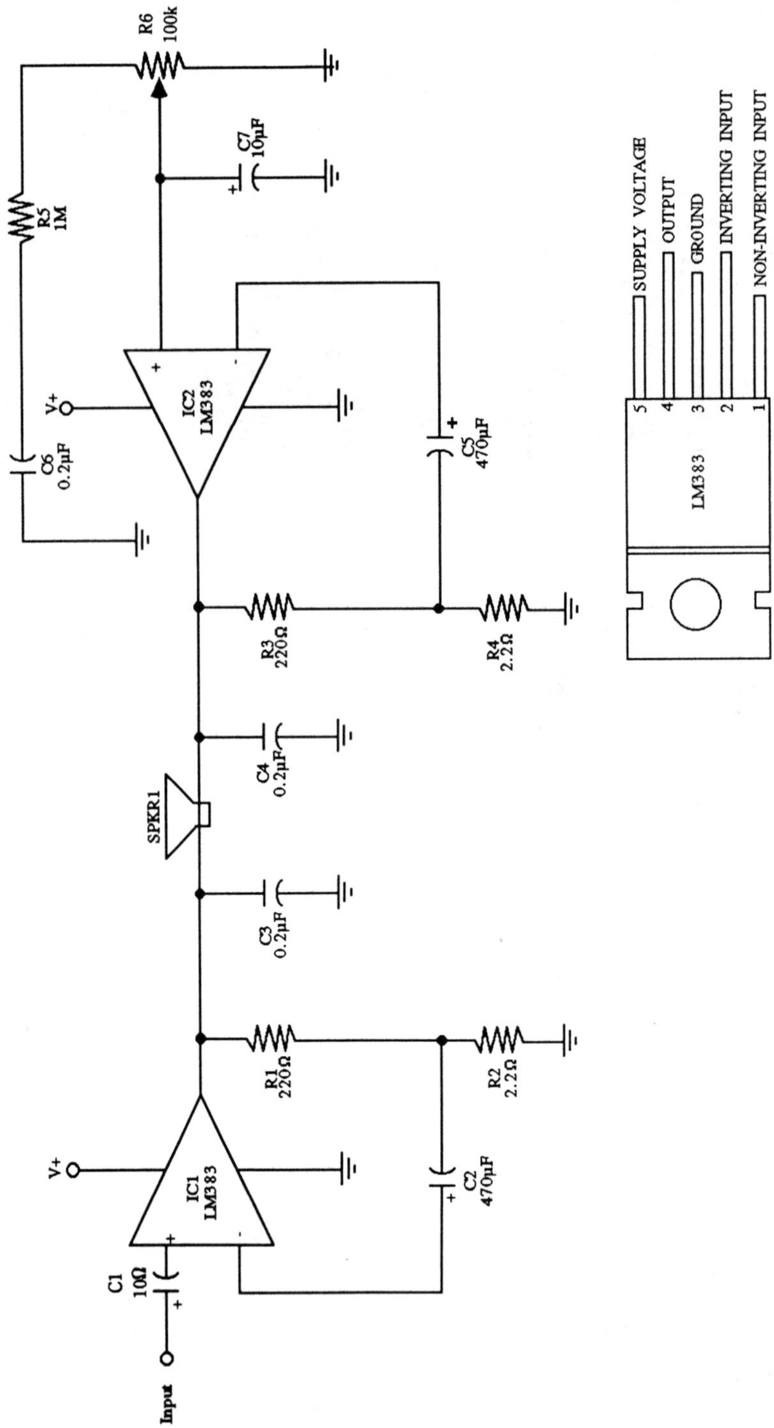

Switch Debouncers

TTL/7400

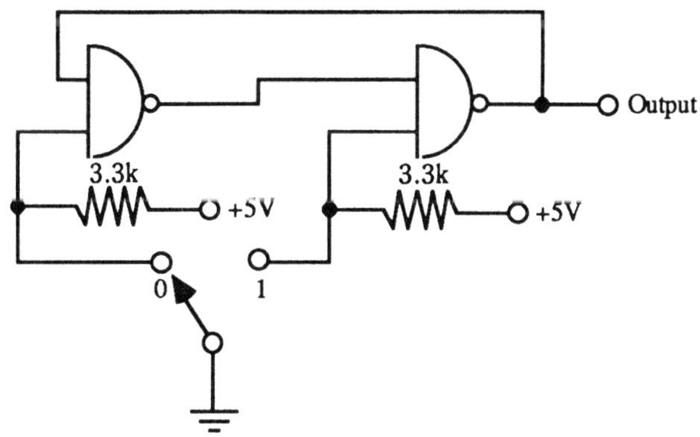

TTL/7402

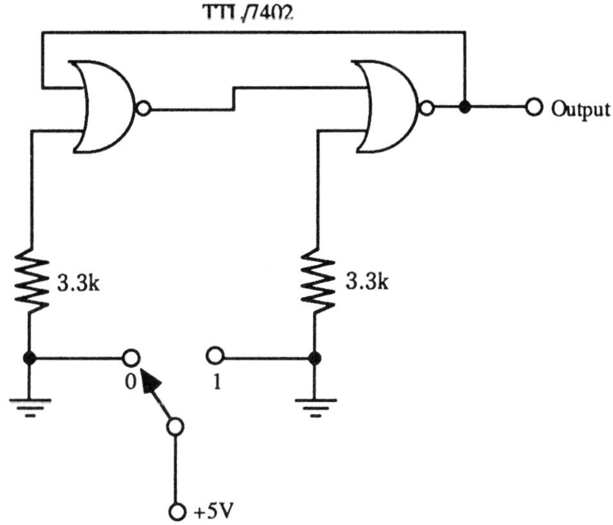

CMOS/4049

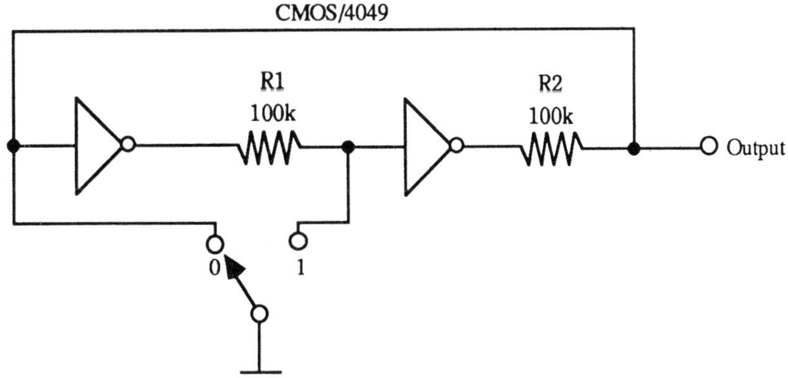

19

Interfacing logic families

MOST INTEGRATED CIRCUITS CAN BE CONNECTED DIRECTLY TO ONE ANOTHER, with no additional components. However, some special design provisions should be made when mixing CMOS and TTL logic families and when interfacing to or from mechanical switches, light-emitting diodes (LEDs), optoisolators, relays, comparator ICs, and operational amplifiers (op amps). Refer to the figures below for more information on interfacing these components.

Circuit Index

TTL- to-CMOS Level Translation

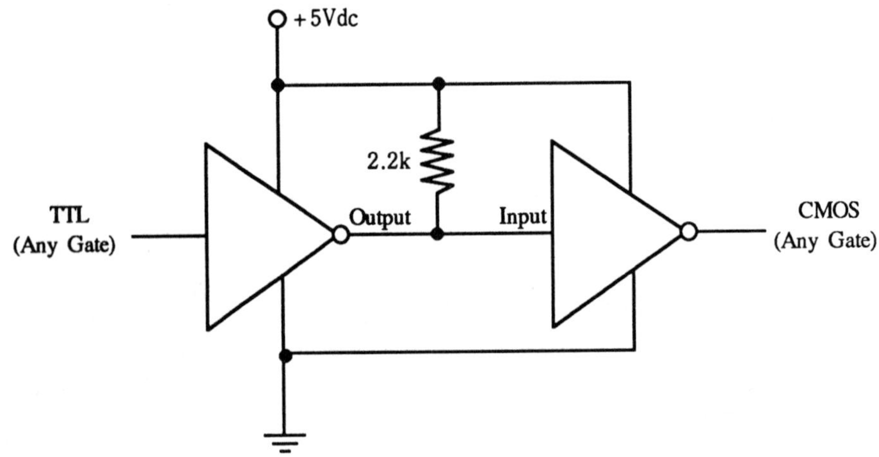

TTL to CMOS (Same Supply Voltage)

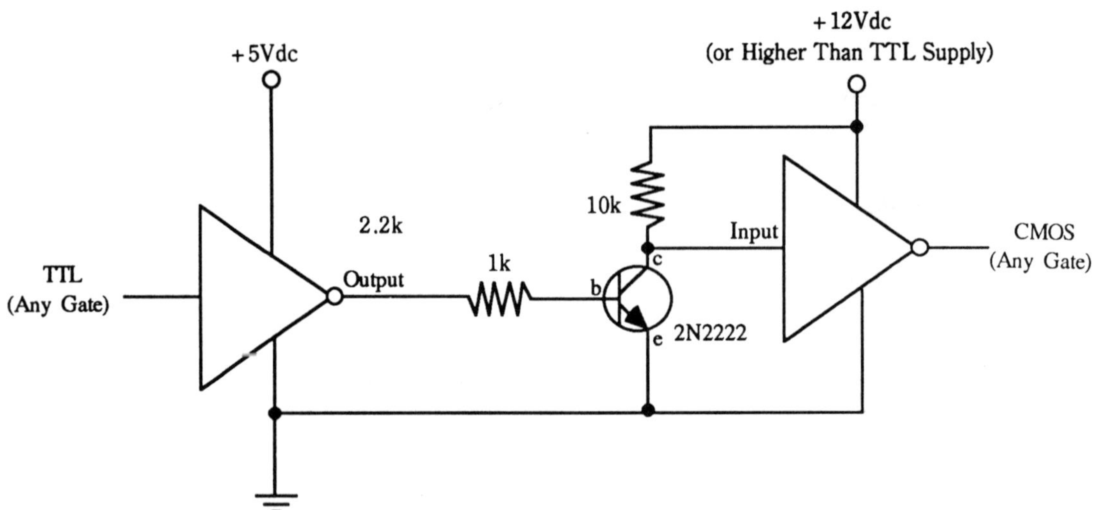

TTL to CMOS (Different Supply Voltage)

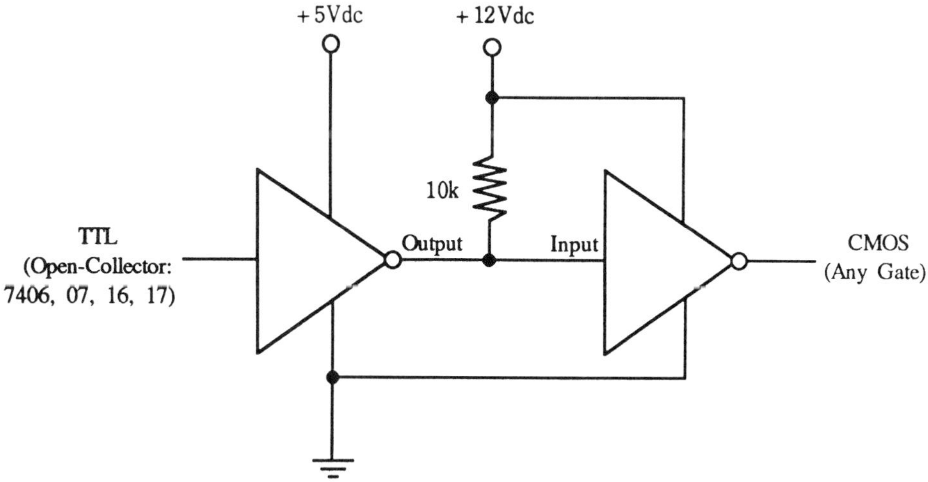

TTL to CMOS (Different Supply Voltage)

CMOS-to-TTL Level Translation

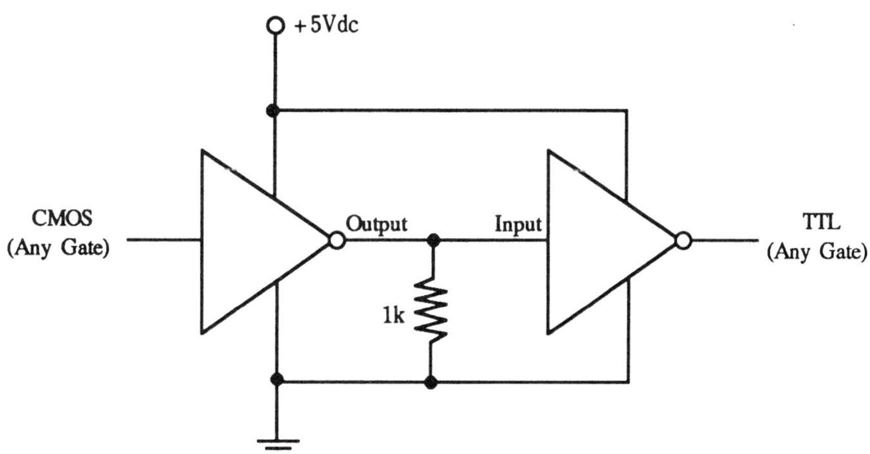

CMOS to TTL (Same Supply Voltage)

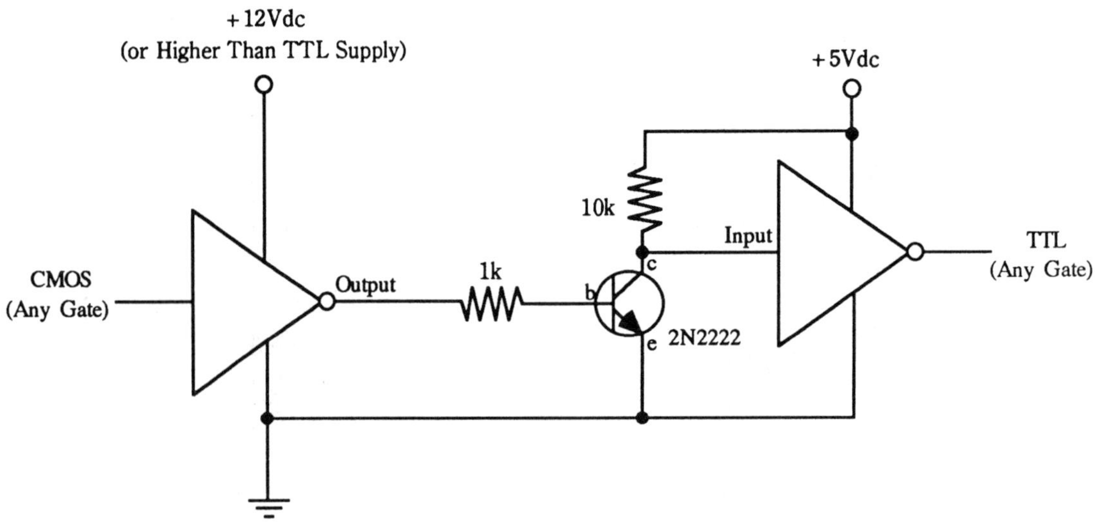

CMOS to TTL (Different Supply Voltage)

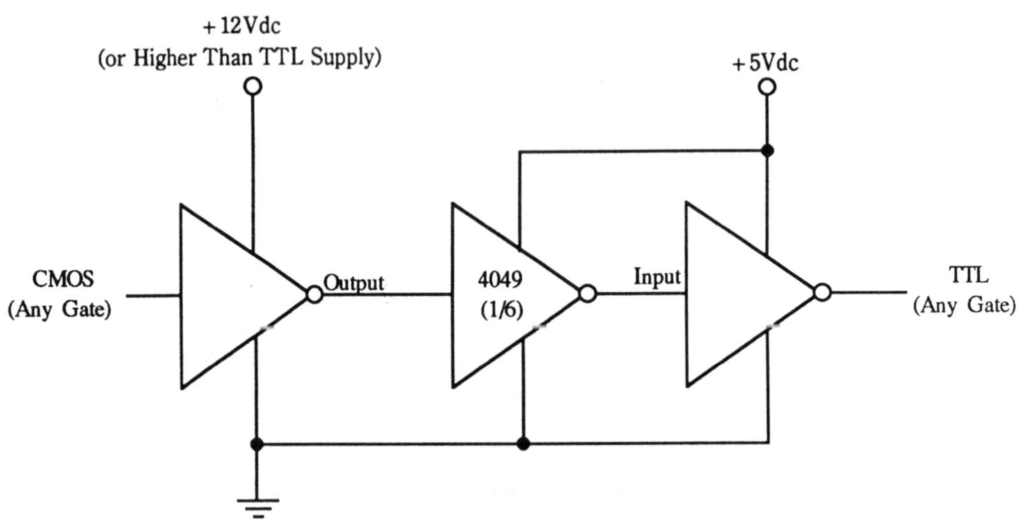

CMOS to TTL (Different Supply Voltage)

Interfacing Op Amps to CMOS and TTL

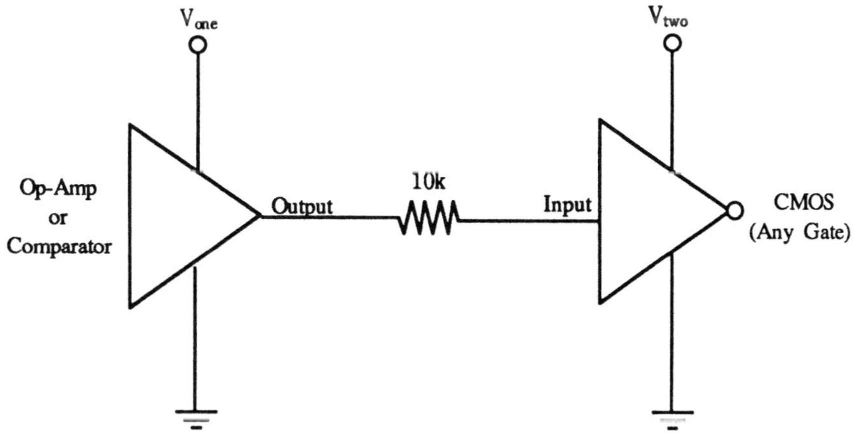

Op-Amp/Comparator to CMOS—$V_{one} = V_{two}$

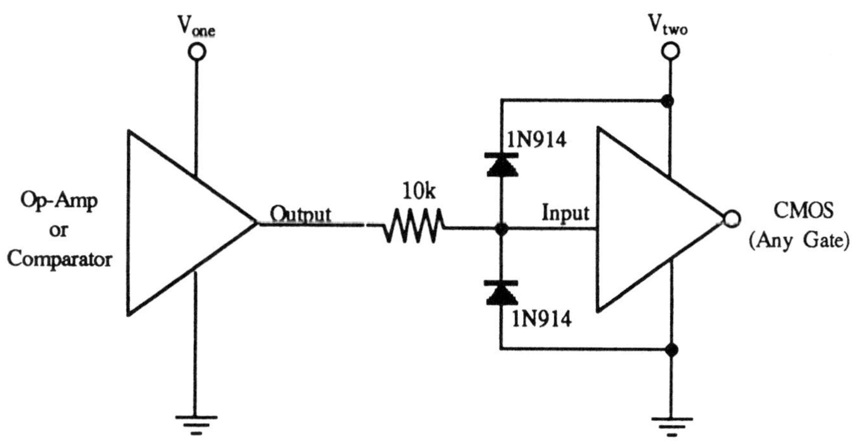

Op-Amp/Comparator to CMOS—$V_{one} = V_{two}$

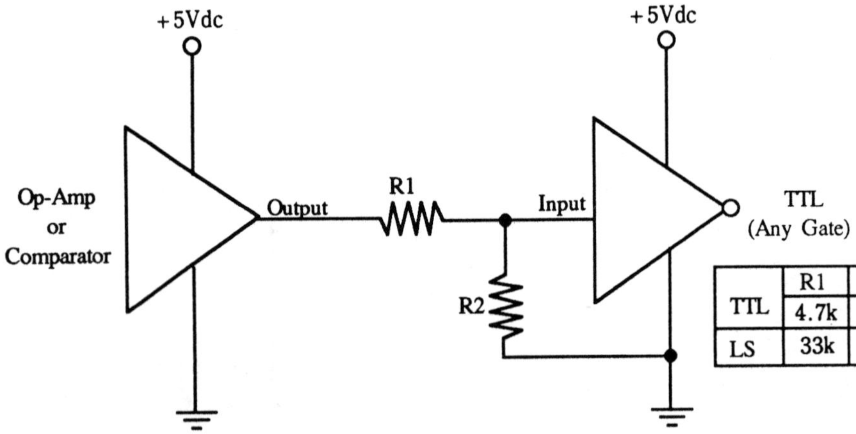

	R1	R2
TTL	4.7k	1.5k
LS	33k	4.7k

Op-Amp/Comparator to TTL—V_{one} = V_{two}

Optoisolator Circuits

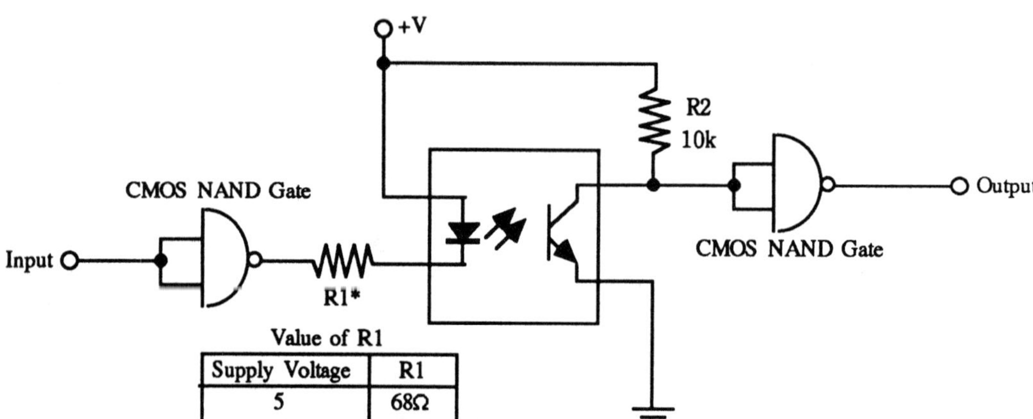

Value of R1

Supply Voltage	R1
5	68Ω
6	100Ω
9	150Ω
12	220Ω

Basic CMOS Optoisolator

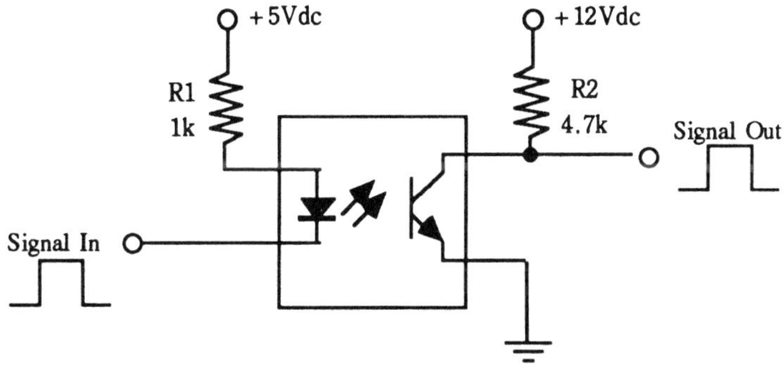

Level Shifer—5Vdc to 12 Vdc
Noninverted Output

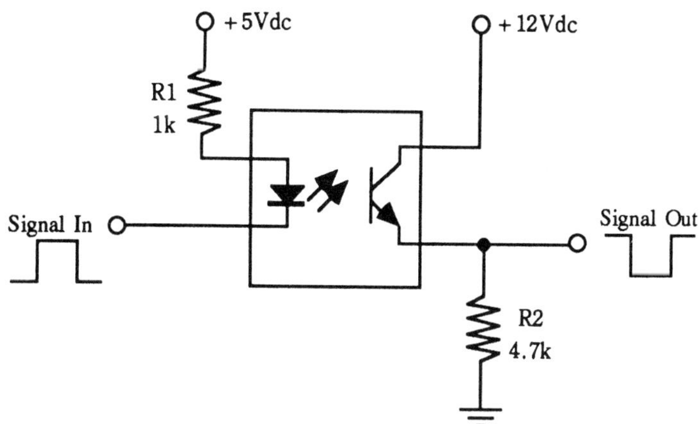

Level Shifer—5Vdc to 12Vdc
Inverted Output

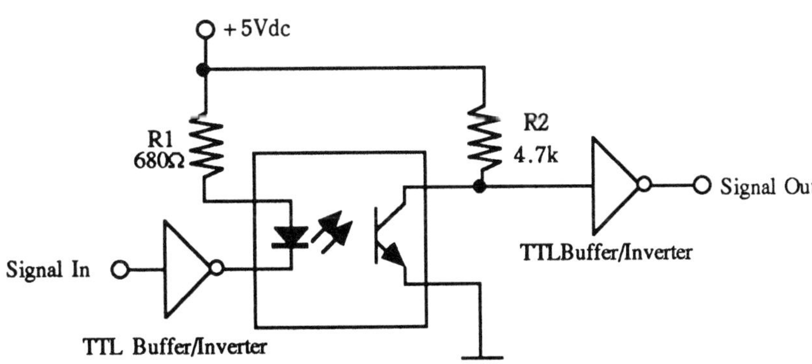

TTL-TTL Isolator

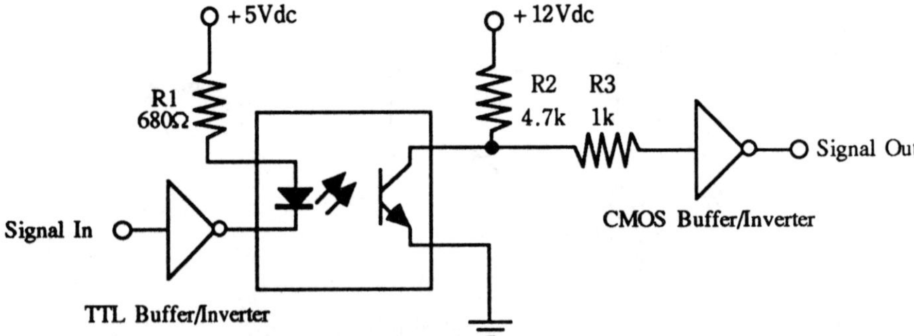

TTL-CMOS Isolator—5Vdc to 12Vdc

CMOS and TTL to Relay and LED Drivers

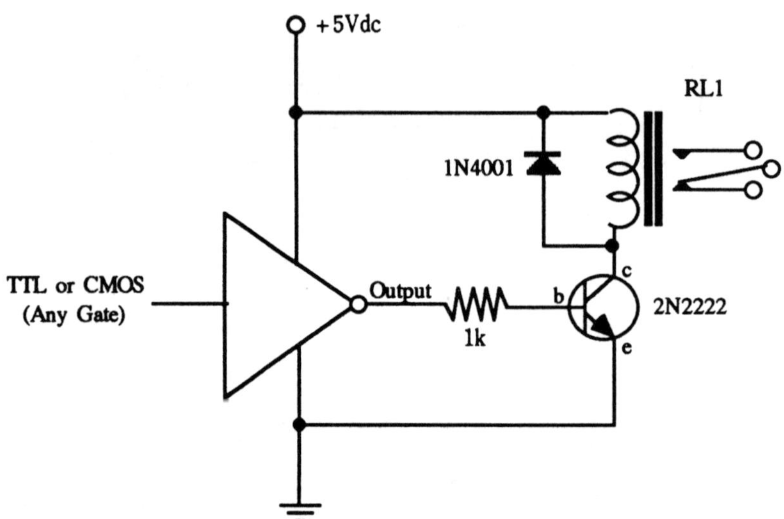

TTL & CMOS to Relay

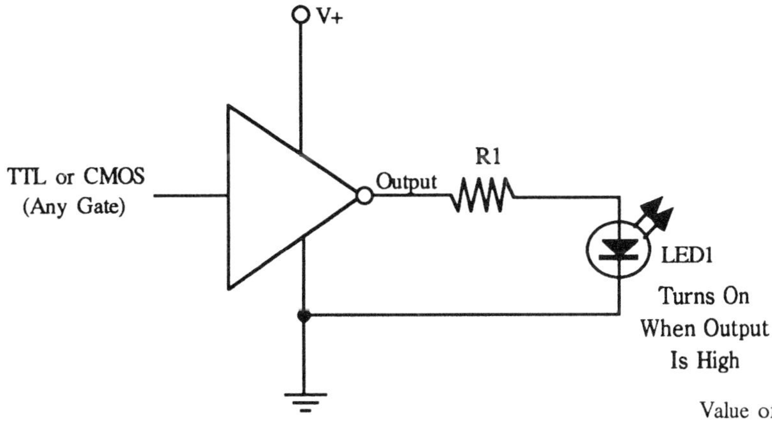

Value of R1

	+5VDC	+12VDC
10mA	330Ω	1K
20mA	180Ω	560Ω
30mA	110Ω	330Ω

Note: 30mA Current May Destroy Some LE

TTL & CMOS to LED

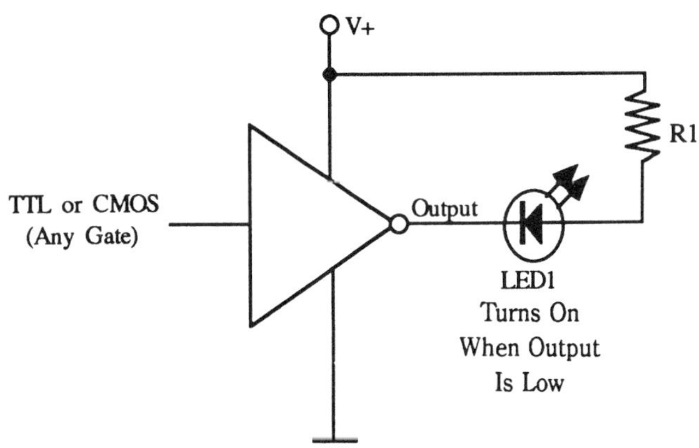

TTL & CMOS to LED

20

Benchtop projects

THIS CHAPTER DETAILS 12 HANDY PROJECTS YOU CAN BUILD FOR YOUR ELECTRONICS test bench. You'll find circuits for power supplies, a logic probe, a logic pulser, a function generator, and even a versatile 10 MHz frequency counter. These projects were selected because of their relative ease in construction and use of low-cost parts.

Circuit Index

Power Supplies

Following are several power supply designs you can use to provide operating juice to your electronic projects. The designs show you how to construct a:

- 5-volt dc regulated power supply
- 12-volt dc regulated power supply
- Quad ±5- and ±12-volt regulated power supply
- Adjustable (3- to 20-volt dc) regulated power supply

Note that the power supplies are similar with the exception of different values for capacitors, diode bridges, and other components. You can alter the schematics to create power supplies that provide other voltage levels. The multivoltage supply is designed to provide the four voltages common in laser support systems: +5 volts, +12 volts, −5 volts, and −12 volts.

Caution: Be extra careful when wiring the power supply circuits, and triple-check your work. Never operate the supply when the top of the cabinet is off unless you are testing it. Even then, stay away from the incoming ac. Touching it can cause a serious shock, and depending on the circumstances, it could kill you. Be extra certain that no wires touch the chassis or front panel. Use an all-plastic enclosure whenever possible. Do not operate the power supply if it has gotten wet, has been dropped, or shows signs of visible damage. Fix any problems *before* plugging it in. Do not assemble the supply without a fuse, and don't use a fuse rated much higher than double the output current. Defeating the fuse protection diminishes or eliminates the only true safety net in power supply circuits.

Single-voltage power supply

Refer to Figs. 20-1 and 20-2 for schematics of the single-voltage power supplies. Figure 20-1 shows the circuit for a +5-volt supply; Fig. 20-2 shows the circuit for a +12-volt supply. There are few differences between them, so the following discussion applies to both. For the sake of simplicity, I refer just to the +5-volt circuit. Parts lists for the two supplies are provided in Tables 20-1 and 20-2.

For safety, the power supply must be enclosed in a plastic or metal chassis (plastic is better because there is less chance of a short circuit). Use a perforated board to secure the components and solder them together using 18- or 16-gauge insulated wire. Do not use point-to-point wiring, where the components are not secured to a board.

Alternatively, you can make your own circuit board using an etching kit. Before constructing the board, collect all the parts and design the board to fit the specific parts you have. There is little size standardization when it comes to power supply components and large-value electrolytic capacitors, so presizing is a must.

When using the bridge rectifier, be sure to connect the leads to the proper

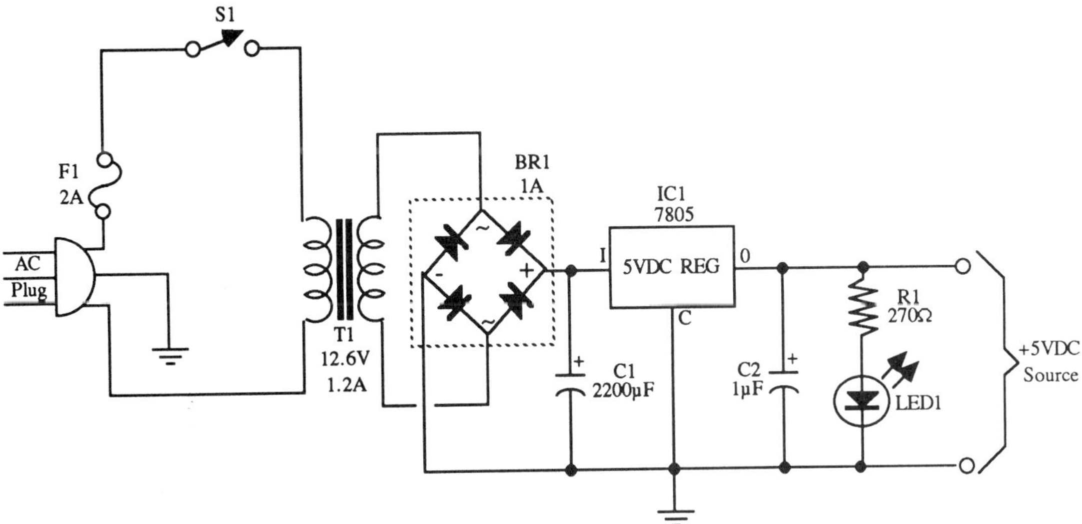

Fig. 20-1. Circuit for the 5-volt power supply.

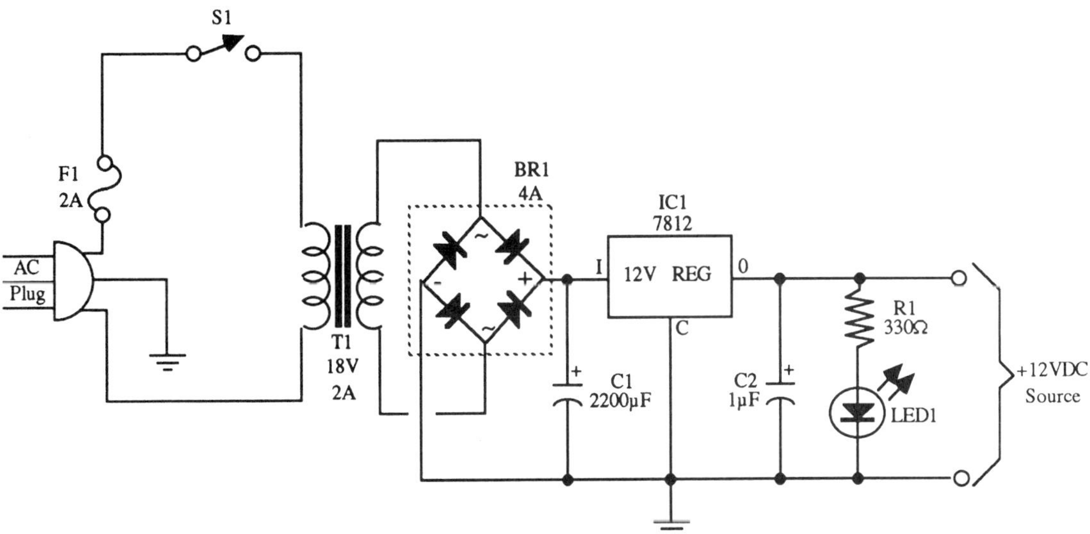

Fig. 20-2. Circuit for the 12-volt power supply.

terminals. The two terminals marked with a wavy line connect to the transformer. The (+) and (−) terminals are the output and must connect as shown in the schematic. A 5-volt, 1-amp regulator (a 7805) is used to maintain the voltage output at a steady 5 volts. The 7805 regulator comes in different styles, shown

Table 20-1. Parts List for 5-Volt Power Supply

IC1	7805 + 5 Vdc voltage regulator
R1	270 Ω resistor
C1	2200 μF electrolytic capacitor
C2	1 μF electrolytic capacitor
BR1	Bridge rectifier, 1 amp
LED1	Light-emitting diode
T1	12.6-volt, 1.2-amp transformer
S1	SPST switch
F1	Fuse (2 amps)
Misc.	Ac plug, cord, fuse holder, cabinet

All resistors are 5 to 10 percent tolerance, 1/4 watt. All capacitors are 10 to 20 percent tolerance, rated at 35 volts or more.

Table 20-2. Parts List for 12-Volt Power Supply

IC1	7812 +12 Vdc voltage regulator
R1	330 Ω resistor
C1	2200 μF electrolytic capacitor
C2	1 μF electrolytic capacitor
BR1	Bridge rectifier, 4 amps
LED1	Light-emitting diode
T1	18-volt, 2-amp transformer
S1	SPST switch
F1	Fuse (2 amps)
Misc.	Ac plug, cord, fuse holder, cabinet

All resistors are 5 to 10 percent tolerance, 1/4 watt. All capacitors are 10 to 20 percent tolerance, rated at 35 volts or more.

in Fig. 20-3:

- The T-suffix regulator comes in a TO-220 transistor-style package and delivers up to 1 amp of current.
- The K-suffix regulator comes in a TO-3 transistor-style package and delivers up to 3 amps of current.

NOTE: Some voltage regulators in the popular 78XX and 79XX series might be available in TO-3 and TO-220 cases with different current delivering capacities. Check the data sheet for the device you are using. In any case, the regulator must be mounted on a suitable heatsink to deliver its full current output.

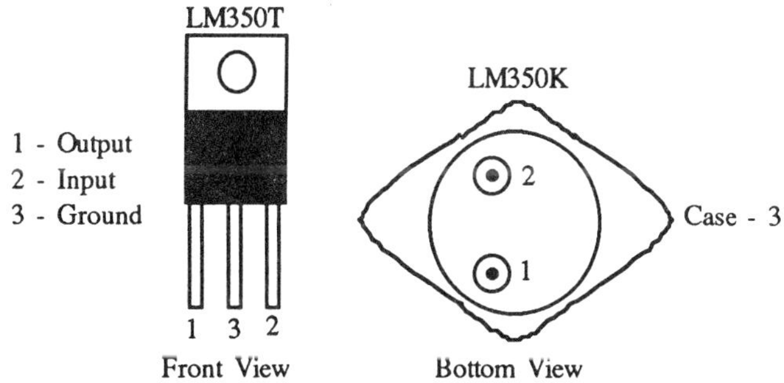

Fig. 20-3. TO-3 and TO-220 packages and pinouts.

You can increase the current rating for your power supply as long as the transformer and bridge rectifier will comply. Be sure the transformer can deliver at least the minimum required current, preferably 50 to 100 percent more. The same rule of thumb applies to the bridge rectifier.

Unless you plan on using the power supply for testing only small circuits, you might want to opt for the larger capacity regulator. You never know when the extra capacity will come in handy.

Capacitors C1 and C2 filter the ripple inherent in the rectified dc found at the outputs of the bridge rectifier. With the capacitors installed as shown (note the polarity), the ripple at the output of the power supply is negligible.

The output terminals are insulated binding posts. Don't leave the output wires bare because they could accidentally touch one another and short the supply. Solder the output wires to the lug on the binding posts, and attach the posts to the front of the power supply chassis. The posts accept bare wires, alligator clips, or even banana plugs.

Multiple-voltage power supply

The multivoltage power supply is like four power supplies in one. Rather than using four bulky transformers, however, this circuit uses just one, tapping the voltage at the proper locations to operate the +5V, +12V, −5V, and −12V regulators.

The circuit, as shown in Fig. 20-4, is really two power supplies in one. One-half of the supply provides +12 and −12 volts; the other half provides +5 and −5 volts. Each side is connected to a common transformer, fuse, switch, and wall plug. See Table 20-3 for a parts list.

The basic difference between the multivoltage supply and the single-voltage supplies described earlier in this chapter is the addition of the negative power regulators. Circuit ground is the center tap of the transformer. A board layout for both +12/−12 and +5/−5 sections is shown in Fig. 20-5. Make two boards, one for each section. That is, one board will be the ±5-volt regulators and the

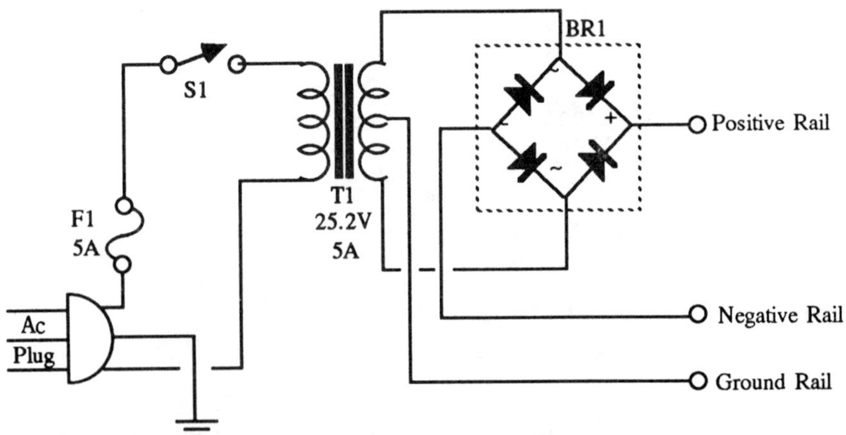

Fig. 20-4. Quad power supply schematic.

Table 20-3. Parts List for Quad Power Supply

IC1	7812 +12 Vdc voltage regulator
IC2	7912 −12 Vdc voltage regulator
IC3	7805 +5 Vdc voltage regulator
IC4	7905 −5 Vdc voltage regulator
C1,C5	2200 μF electrolytic capacitors
C2,C3,C6,C7 C10,C11,C14,C15	1 μF electrolytic capacitors
C4,C8 C12,C16	100 μF electrolytic capacitors
C9,C13	1000 μF electrolytic capacitors
C1,C5	2200 μF electrolytic capacitors

All capacitors are 10 to 20 percent tolerance, rated 35 volts or more.

other board will contain the ±12-volt regulators. The supply provides approximately 1 amp for each of the outputs.
and the other board will contain the ±12-volt regulators. The supply provides approximately 1 amp for each of the outputs.

Use nylon binding posts for the five outputs (ground, +5 V, + 12 V, −5 V, −12 V). Clearly label each post so you don't mix them up when using the supply. Check for proper operation with your volt-ohmmeter.

Adjustable voltage power supply

The adjustable power supply presented below uses an LM317 adjustable voltage regulator. With the addition of a few components, you can select any voltage be-

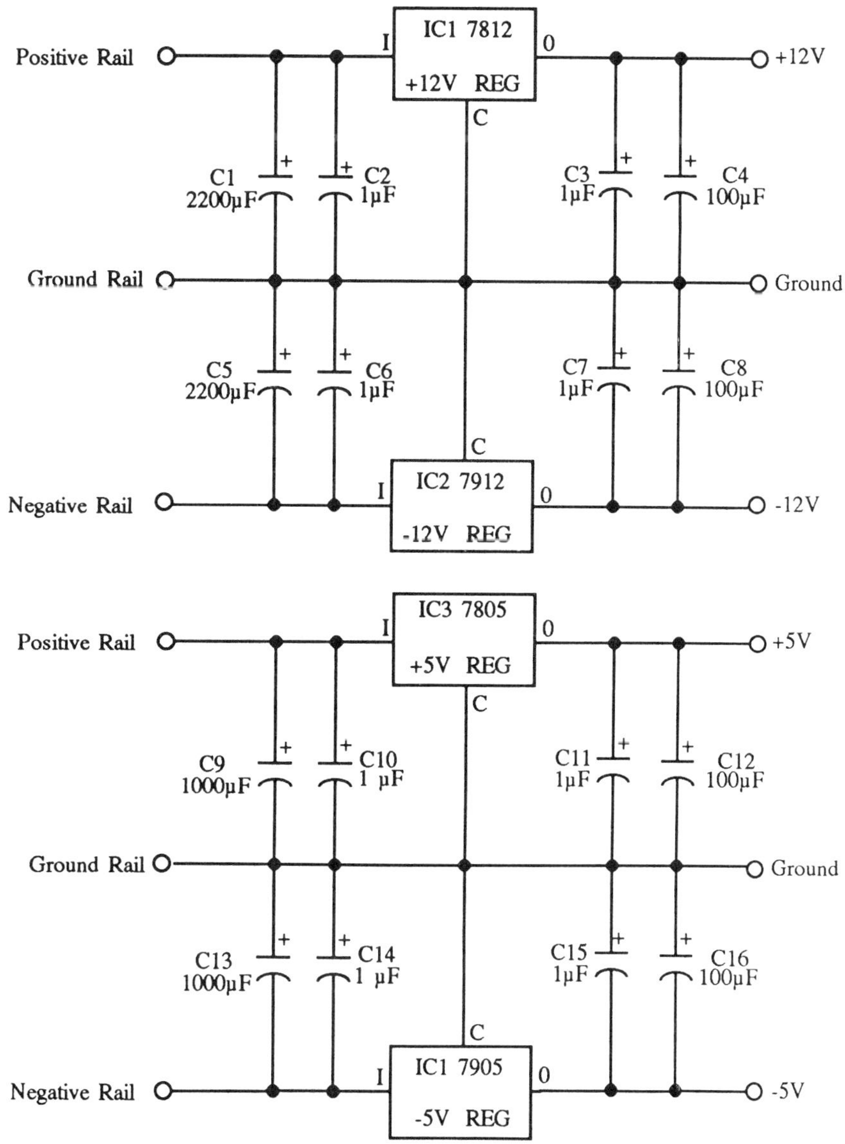

Fig. 20-5. Use this circuit board layout when building the quad power supply.

tween 1.5 to 37 volts. By using a potentiometer, you can select the voltage you want by turning a knob.

The circuit shown in Fig. 20-6 is a no-frills application of the LM317, but it has everything you need to build a well-regulated, continuously adjustable, positive voltage power supply. Table 20-4 provides the parts list. The regulator is rated at over 3 amps, but you must mount it on a heavy-duty heatsink. Although you don't need to forcibly cool the regulator and heatsink, it might be a good idea

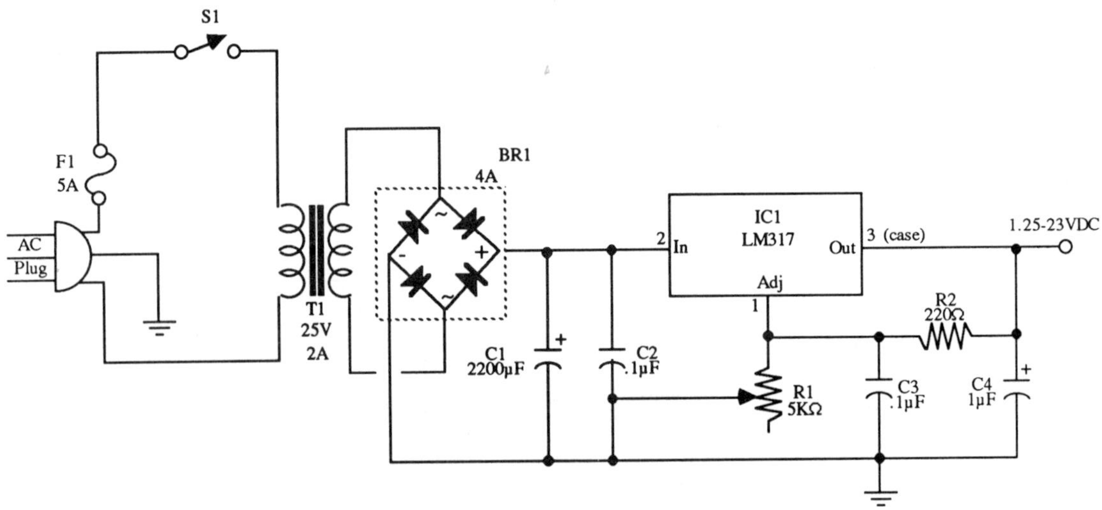

Fig. 20-6. Schematic for a versatile adjustable power supply.

Table 20-4. Parts List for Adjustable Power Supply

IC1	LM317 adjustable positive voltage regulator
R1	5 kΩ potentiometer
R2	220 Ω resistor
C1	2200 µF electrolytic capacitor
C2,C3	0.1 µ disc capacitors
C4	1 µF electrolytic capacitor
BR1	Bridge rectifier, 4 amps
T1	25-volt, 2-amp (or more) transformer
S1	SPST switch
F1	5-amp fuse
Misc.	Ac plug, cord, fuse holder, cabinet

All resistors are 5 to 10 percent tolerance, 1/4 watt. All capacitors are 10 to 20 percent tolerance, rated at 35 volts or more.

to mount it on the outside of the power supply cabinet. Place the regulator and heatsink on the top or back.

Remember that the case of the regulator is the output, so be sure to provide electrical insulation from the heatsink, or a short circuit could result. Use a TO-3 transistor mounting and insulator kit. It has all the hardware and insulating washers you need. Apply silicone grease to the bottom of the regulator to aid in heat transfer.

For greatest flexibility, you can build two adjustable power supplies and mount them in the same cabinet (use one master transformer if it can deliver enough current—at least 6 amps worth).

Battery-Pack Regulators

Voltage regulators can also be used with battery packs for portable equipment. A 5-volt regulator can be used with a single 6-volt battery to provide a steady supply of 5 volts. The schematic in Fig. 20-7 shows how to connect the parts. Refer to Table 20-5 for a parts list.

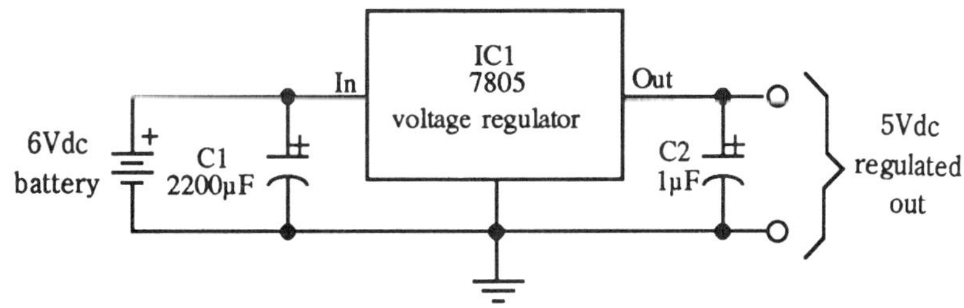

Fig. 20-7. A battery regulator.

Table 20-5. Parts List 5-Volt Battery Regulator

IC1	7805 +5 Vdc voltage regulator
C1	2200 μF electrolytic capacitor
C2	1 μF electrolytic capacitor

All capacitors are 10 to 20 percent tolerance, rated at 35 volts or more.

Battery Monitors

A battery monitor simply provides an aural or visual indicator that a battery is either delivering too much or too little voltage. Figure 20-8 shows a schematic for a simple "window comparator" battery monitor (see Table 20-6 for a parts list). It is designed to be used with 12-volt batteries, but you can substitute one or more of the zener diodes for use with other voltages.

In normal operation, LED1 glows when the voltage from the battery is at least 10 volts. It is also desirable to know if the battery is delivering too much voltage, so a second zener diode is used. If LED2 is on, the circuit is receiving too much power and could be damaged. More likely, however, the battery level will drop, and LED1 will grow dim or flicker off completely. If LED1 is not lit or is dim, the battery needs to be recharged.

Figure 20-9 shows another simple battery monitor with a one LED output. Potentiometer R1 sets the trip point of the circuit, or the minimum voltage level before the circuit is activated. The LED lights when the battery drops below the trip point.

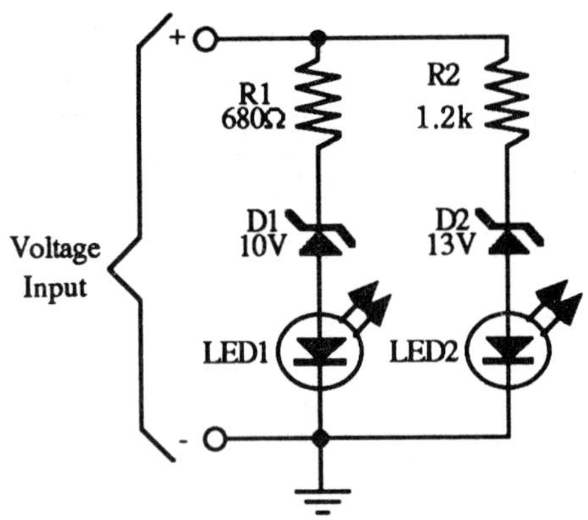

Fig. 20-8. Window comparator battery monitor.

Table 20-6. Parts List for Window Comparator Battery Monitor

IC1	LM339 comparator IC
R1,R2,R3	10 kΩ potentiometers
R4,R5	22 kΩ resistors
R6	47 kΩ resistor
R7	68 kΩ resistor
R8 – R10	330 Ω resistors
LED1 – 3	Light-emitting diodes

All resistors are 5 to 10 percent tolerance, 1/4 watt.

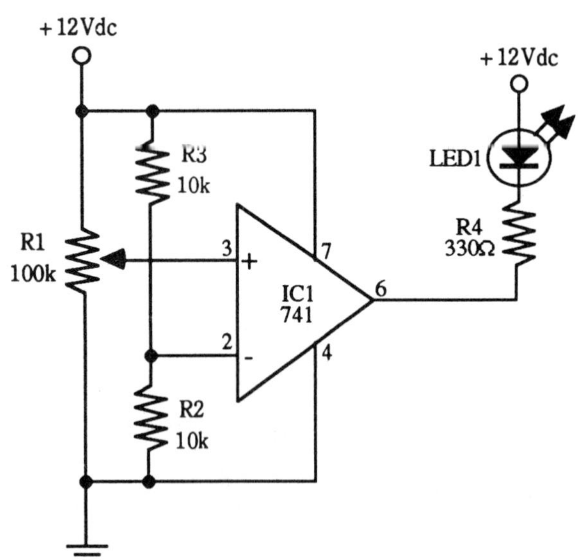

Fig. 20-9. Alternate battery monitor #1.

A similar circuit is shown in Fig. 20-10. Once again, a variable resistor, R1, sets the trip point of the circuit. When the input voltage drops below the trip point, the LED turns on. This circuit can be used with other supply voltages by changing the value of the zener diode and by readjusting R1. Parts lists for these two circuits are provided in Tables 20-7 and 20-8.

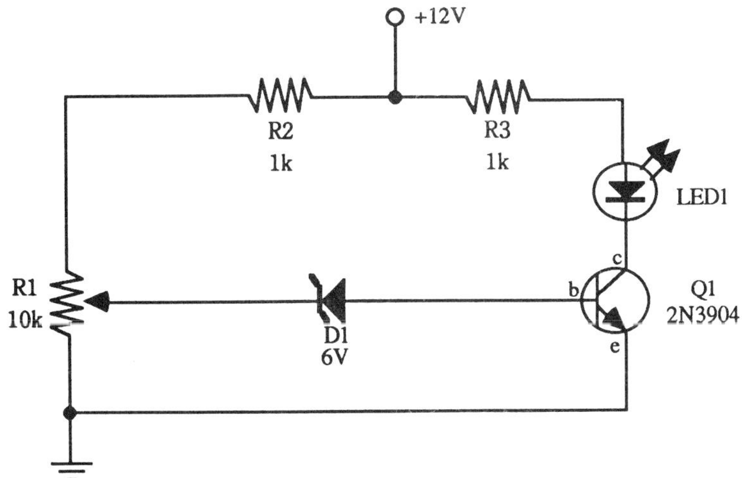

Fig. 20-10. Alternate battery monitor #2.

Table 20-7. Parts List for Alternate Battery Monitor #1

R1	680 kΩ resistor
R2	1.2 kΩ resistor
D1	10-volt zener
D2	13-volt zener
LED1,2	Light-emitting diodes

All resistors are 5 to 10 percent tolerance, 1/4 watt.

Table 20-8. Parts List for Alternate Battery Monitor #2

Q1	2N3904 signal transistor
R1	10 kΩ potentiometer
R2,R3	1 kΩ resistors
LED1	Light-emitting diode
D1	6-volt zener

All resistors are 5 to 10 percent tolerance, 1/4 watt.

Infrared Detector

A number of electronic projects, such a lightwave communicators and see-in-the-dark night scopes, use infrared illumination. The infrared light source, whether it's an LED, infrared spotlight, or other device, emits light in the region above normal human vision, typically 780 to 900 nanometers in the light spectrum. Because the light is invisible to your eyes, it can be difficult to test the operation of the infrared light source.

An infrared detector, discussed below, senses the presence of infrared light and provides you with a visual indication. The sensor is easy to build and inexpensive. It can be used to detect the presence of all types of infrared light including that from laser diodes in compact disc players, video disc players, and VCRs (videocassette recorders use infrared LEDs in tape threading and transport mechanisms).

Circuit description

The infrared light sensor is a simple circuit that employs only three parts: an infrared phototransistor, an LED, and a resistor. The schematic is shown in Fig. 20-11. A parts list is provided in Table 20-9. All the parts are commonly available at Radio Shack and most other electronic parts stores. You need some sort of battery supply to operate the circuit. I built the circuit with four AA batteries enclosed in a self-contained battery compartment. The compartment also serves as the case for the sensor electronics.

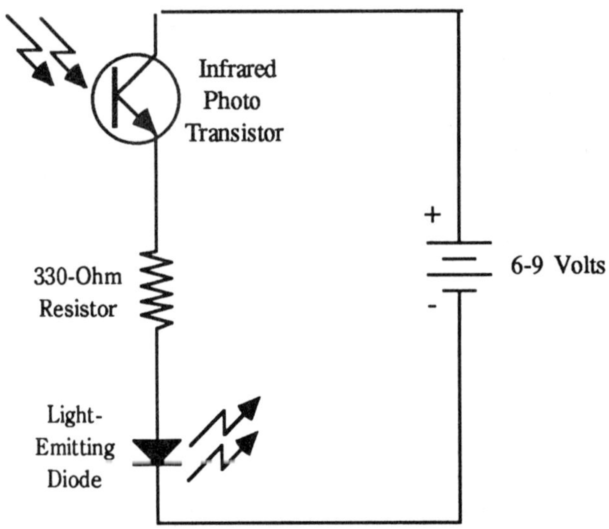

Fig. 20-11. Infrared light detector schematic.

Table 20-9. Parts List for Infrared Detector

Q1	Infrared photo transistor (almost anyone will do)
R1	330 Ω resistor, 1/4 watt
LED1	Light-emitting diode
Misc.	6- to 9-volt battery, battery clip, project box.

Alternatively, you can use a 9-volt transistor battery mounted in a simple clip holder. The resistor limits the current flowing through the LED, but the 330-ohm value selected is safe for use with supply voltages up to 9 volts. If you use a higher supply voltage, increase the value of the resistor, using Ohm's law. Failure to do this could cause the visible LED in the circuit to burn out.

You can determine the value of the resistor by taking the supply voltage and subtracting the voltage drop through the LED (usually 1.2 volts). Divide the result by the current you want flowing through the LED (usually about 20 milliamps).

For example, with a supply voltage of 9 volts, subtracting the drop in the LED makes 7.8 volts. Divide that by 0.02 (20 milliamps) and you get 390. The exact value of resistor to use, therefore, is 390 ohms. There is a lot of slack built into this because LEDs can take somewhat higher currents without risk of damage. On a practical level, don't exceed 40 to 45 milliamps.

For best results, solder the components on a small perforated board, or use an 8-pin IC socket. The components fit in the contacts of the socket without soldering, but you might want to solder the leads in place to prevent them from coming out. Soldering into the socket is a little tricky because the plastic can melt. An easy way is to fit the component leads into the contacts of the socket, and briefly touch the leads with the tip of the soldering iron. Before the plastic of the socket melts, apply the solder. The electrical contact has already been made inside the socket by the contacts; the solder simply serves to anchor the leads in place.

Place the assembly in a box, like the battery holder shown in Fig. 20-12, to prevent damage during storage and use. The LED is mounted on the other side of the socket so that it can be seen when the sensor is pointed towards the LED under test.

Logic Probe

The logic probe is the staple of the digital circuit experimenter. Logic signals can change state quickly—too quickly to be registered on a volt-ohmmeter. The circuit in Fig. 20-13 has three LEDs to show three logic states: high, low, and pulsing. The circuit consists of just two common ICs and associated parts, as detailed in the parts list in Table 20-10.

Most logic probes are built for handheld use, typically housing the components into a slender pen-like housing. If you like this approach, build your logic probe as shown in Fig. 20-14, and place the board in a toothbrush holder or some other suitable enclosure. The probe tip is a 3d nail. You can also enclose the components in a small box, as shown in Fig. 20-15, and attach leads to alligator clips and a handheld probe tip.

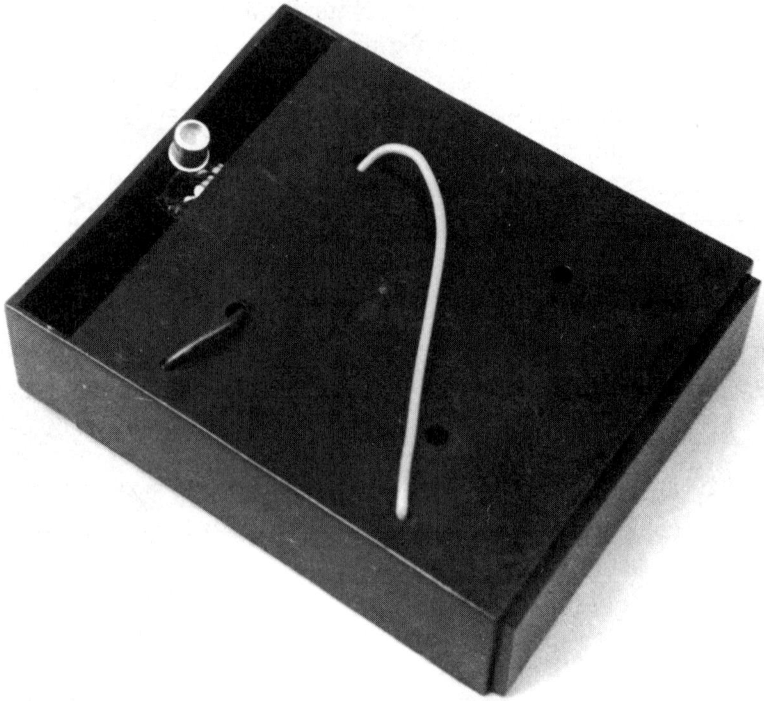

Fig. 20-12. The infrared light detector, completed.

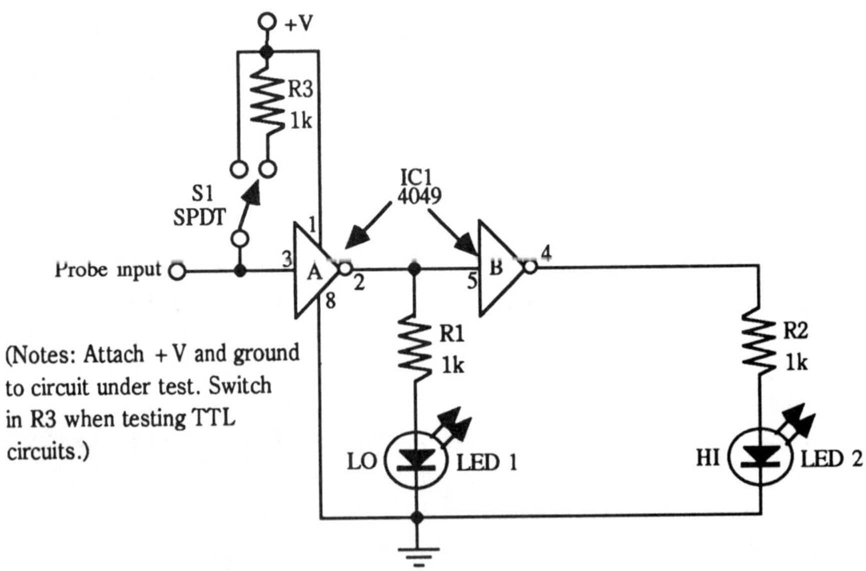

Fig. 20-13. Schematic for the logic probe.

Table 20-10. Parts List for Logic Probe

IC1	4049 CMOS hex inverter IC
R1,R2,R3	1 kΩ resistors
LED1	Light-emitting diode, green
LED2	Light-emitting diode, red
S1	SPDT switch

All resistors are 5 to 10 percent tolerance, 1/4 watt.

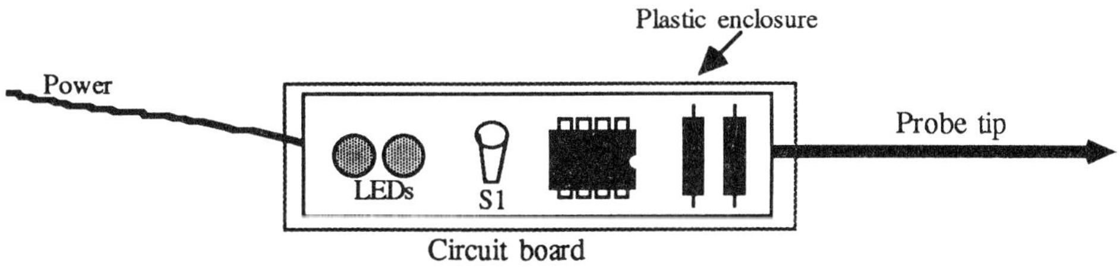

Fig. 20-14. Pen-like arrangement for the logic probe.

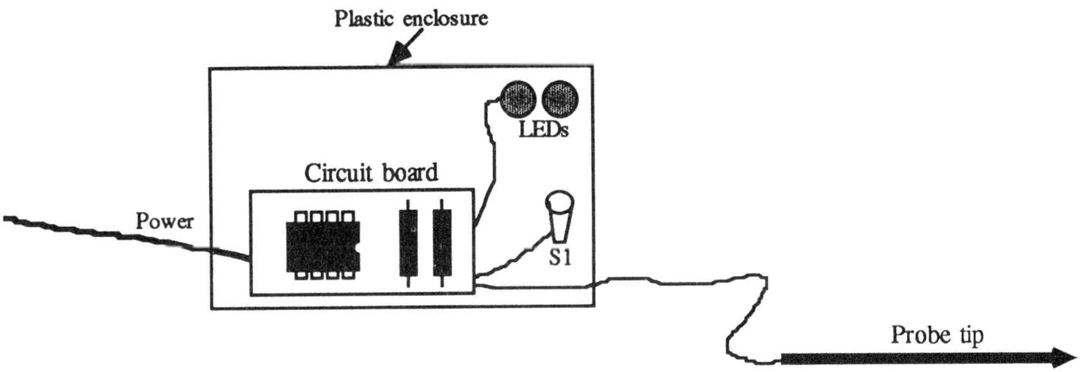

Fig. 20-15. Separate box arrangement for the logic probe.

To use your logic probe, connect the power leads to the circuit under test (the circuit should not use voltages exceeding ±18 Vdc). Touch the tip of the probe to the test point. If the logic level is high, LED 2 (green) should glow. If the logic level is low, LED 1 (red) should glow. If the logic state is toggling quickly between high and low, both LEDs should glow. If no LED glows, the circuit is either not working properly, or there is no voltage at the test point.

Logic Pulser

A logic pulser provides a train of square waves to a circuit under test. You can use a logic pulser to simulate activity in a circuit or to force a particular component in a circuit to change logic states.

Figure 20-16 shows a circuit for a simple logic pulser (see the parts list in Table 20-11). The pulser is set to deliver a steady stream of high/low pulses at a rate of approximately 500 Hz. A momentary switch lets you apply a steady high state to the circuit under test. Note that the momentary switch employs contact "debouncing" circuitry to provide a clean level change to the circuit under test.

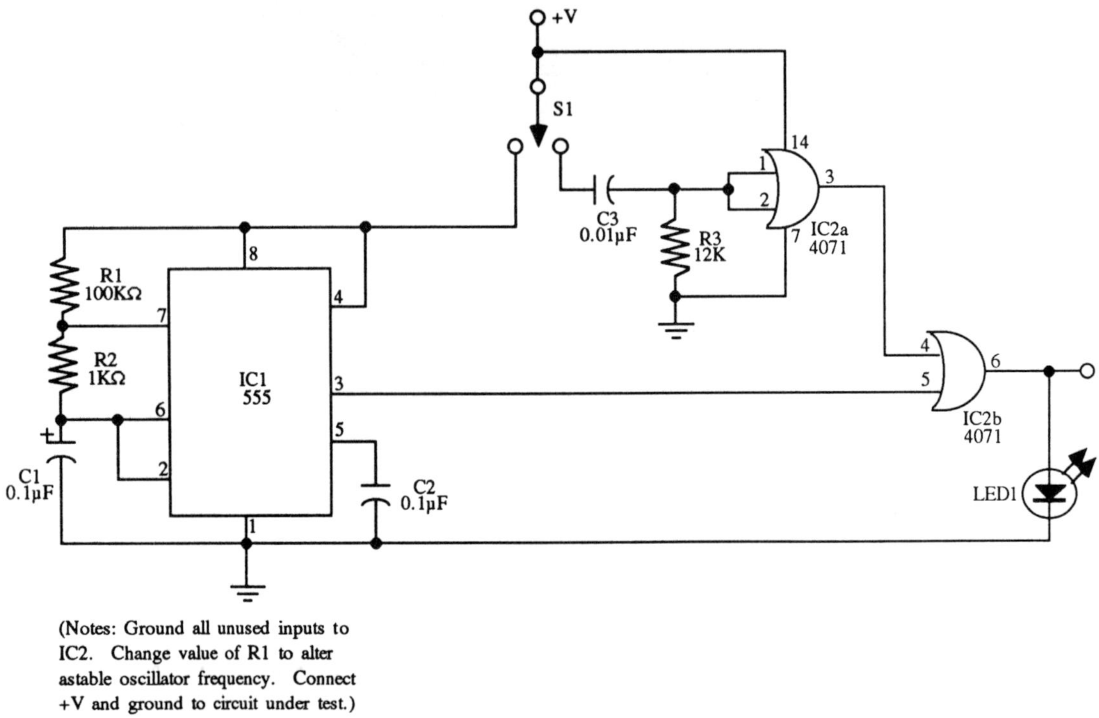

Fig. 20-16. The logic pulser circuit.

As with the logic probe, detailed above, pack the components into a suitable enclosure, either some type of handheld pen-like case, or a small hobby box. Note S1 in the parts list and schematic. It is a double-throw (on-off-on) combination toggle where one of the ON positions is spring loaded. Push the switch one direction and the toggle stays there; push it the other direction and it springs back. This type of double-throw momentary toggle switch can be found at most electronics stores. Make sure that the switch you get has a center-off position.

To use the circuit, attach the power leads to the circuit under test (but in no

Table 20-11. Parts List for Logic Pulser

IC1	555 timer IC
IC2	4071 CMOS AND gate IC
R1	100 kΩ resistor
R2,R4	1 kΩ resistors
R3	12 kΩ resistor
C1,C2	0.1 μF disc capacitors
C3	0.01 μF disc capacitor
LED1	Light-emitting diode
S1	SPDT switch

All resistors are 5 to 10 percent tolerance, 1/4 watt. All capacitors are 10 to 20 percent tolerance, rated at 35 volts or more.

case more than +18 Vdc). Touch the probe tip to the test point. To activate the 500 Hz pulses, touch the PULSE switch (push one way for continuous operation; push the other way for momentary operation). To force the output high, press the HIGH button.

Slow-Rate Variable Pulser

The slow-rate variable pulser is a modified and simplified version of the logic pulser, described in the previous section. The slow-rate variable pulser provides a steady stream of pulses from about 3 Hz to 30 Hz, via potentiometer R1. Adjust R1 for the pulse rate you want. An LED provides a visual indication of the pulses. Refer to the schematic in Fig. 20-17 for building the slow-rate variable pulser. The parts list is shown in Table 20-12. A complete pulser is shown in Fig. 20-18.

Use the slow-rate variable pulser in the same fashion as the logic pulser. That is, connect the pulser to the circuit under test (make sure the voltage is under +18 Vdc). Then attach a lead between the output of the pulser and the test point on your circuit. Note that I've omitted an enclosure for my pulser. The simplicity of the circuit (it was built on a solder breadboard, by the way) just doesn't seem to warrant it. Of course, you can enclose yours in a suitable case if you desire.

Function Generator

The function generator shown in the schematic in Fig. 20-19 provides three waveforms: square, triangle, and sine. The circuit uses the versatile Exar XR-2206 single-chip function generator IC. Although it might not look like it at first,

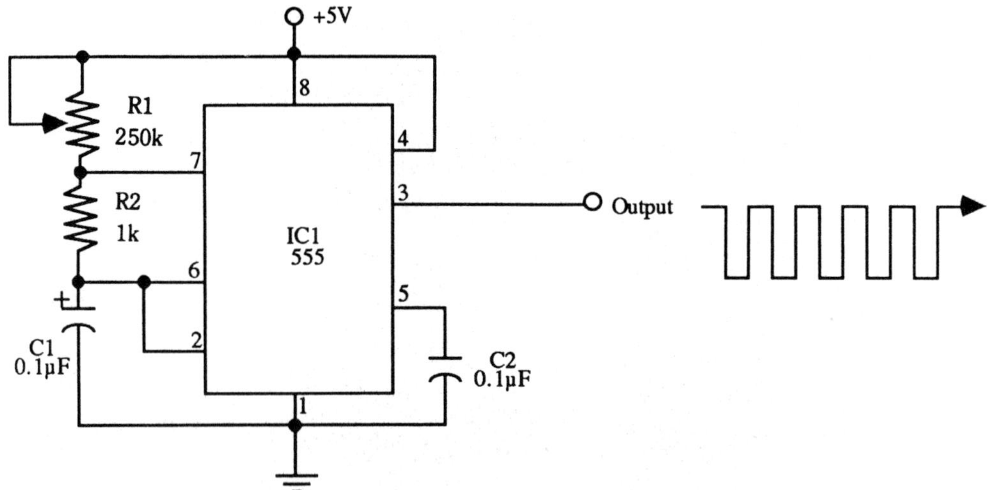

Fig. 20-17. Slow-rate variable pulser circuit.

Table 20-12. Parts List for Slow-Rate Logic Pulser

IC1	555 timer IC
R1	250 kΩ potentiometer
R2	1 kΩ resistor
C1,C2	0.1 µF disc capacitors

All resistors are 5 to 10 percent tolerance, 1/4 watt. All capacitors are 10 to 20 percent tolerance, rated at 35 volts or more.

this chip—along with an assortment of resistors, capacitors, and potentiometers—is all you need to make a fully serviceable function generator. With the parts shown (and detailed in the parts list in Table 20-13), the circuit provides frequencies from 1 Hz to 100 kHz, in four separate bands, as set by switch S2 (see Table 20-14).

Build the circuit using a universal solderboard or construct your own printed circuit board. House the finished circuit in a suitable enclosure. Note that only switches S1 and S2 and potentiometers R12 and R13 are mounted externally; all other potentiometers are mounted on the circuit board. Output and input terminals are BNC connectors. After assembly is complete, dial potentiometers R2, R10, and R11 to their middle positions.

The function generator provides two separate outputs. The TRI/SINE output provides triangle or sine waves, determined by the setting of switch S2. Set in one position, the output is triangle waves; set in the other position the output is

Fig. 20-18. Finished slow-rate pulser.

sine waves. The SQUARE output, which is always enabled, provides a steady train of square pulses.

To use the function generator, connect the TRI/SINE or SQUARE output to the circuit under development. If you are using the TRI/SINE output, flip S2 to TRI or SINE as desired. Turn the AMPLITUDE control to adjust the strength of the signal and the FREQUENCY control to adjust the repetition rate. Use switch S1 to dial in the frequency range that you want.

If you have an oscilloscope, you can adjust R2 (dc offset) and R10/R11 (distortion) to "tweak" the function generator for better output response. Connect the TRI/SINE output of the function generator to the oscilloscope and flip S2 to SINE (closed). Dial the AMPLITUDE knob so that you see a nonclipped (not distorted) waveform. Adjust R10 and R11 alternately to reduce any distortion that might be present.

Now turn to R2, dc offset. Look at the sine wave displayed on the oscilloscope tube and turn R2 so that the wave is equally balanced about the 0-volt mark. The dc offset is set properly when the sine wave deviates from 0 volts by the same amount both positive and negative; in other words, *exactly* +5 Vdc and *exactly* −5 Vdc.

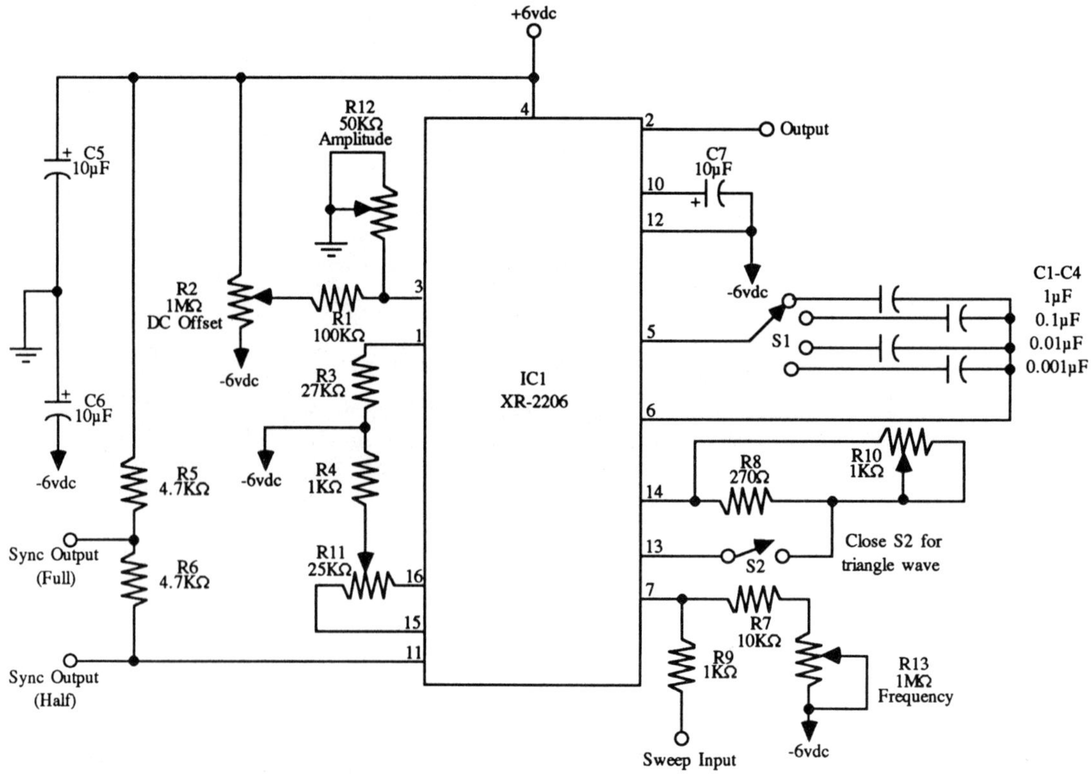

Fig. 20-19. Schematic for the function generator.

Table 20-13. Parts List for Function Generator

R1	100 kΩ
R2,R13	1 MΩ potentiometer
R3	27 kΩ
R4,R9	1 kΩ
R5,R6	4.7 kΩ
R7	10 kΩ
R8	270 kΩ
R10	1 kΩ potentiometer
R11	25 kΩ potentiometer
R12	50 kΩ potentiometer
C1	1 μF disc capacitor
C3	0.01 μF disc capacitor
C4	0.001 μF dis capacitor
C5,C6,C7	10 μF polarized electrolytic capacitor
S1	SP4T rotary switch
S2	SPST switch

IC1	Exar XR-2206 function generator IC
Misc.	Project box, 6 Vdc power supply, input jacks, output jacks, etc.

All resistors are 5 to 10 percent tolerance. All capacitors are 10 to 20 percent tolerance, rated at 35 volts or more.

Table 20-14. Setting Via S2

Switch Setting	Frequency Range
C = 1 μF	10 Hz – 100 Hz
C = 0.1 μF	100 Hz – 1 kHz
C = 0.01 μF	1 kHz – 10 kHz
C = 0.001 μF	10 kHz – 100 kHz

10 MHz Frequency Counter

A frequency counter is a handy piece of test gear. Logic probes can indicate when a circuit is oscillating, but they can't report the frequency of that operation. A scope can be used to determine signal frequency, but it requires interpretation of the test results. The frequency counter does it all, almost instantaneously.

Figure 20-20 shows an easy-to-build frequency counter based around the Intersil ICM7208 seven-digit frequency-counter chip. This chip, which costs about $20 from mail order (one source is Circuit Specialists; see the address in appendix A, "Sources"), provides almost all the smarts you need to build a complete, working test instrument. A few extra components are required to shape the incoming signal and provide an oscillator controller. Seven common-cathode seven-segment displays are used to indicate the frequency. The chip handles frequencies from 1 Hz to 10 MHz. Table 20-15 provides a parts list.

Build the frequency counter in a suitable case. The frequency counter requires a +5 Vdc power supply, providing a minimum of 750 mA. You can use the 5 Vdc power supply provided earlier in this chapter, or build the circuit into the case of your frequency counter.

Figure 20-21 shows how to wire the seven-segment displays for interfacing to the 7208 chip. Note that not all seven-segment displays use the same pinout arrangement, so check the data sheet that came with yours and make the necessary wiring adjustments. The displays used in the illustration are Radio Shack common-cathode seven-segment LEDs, catalog number 276-075. You can use these or anything equivalent.

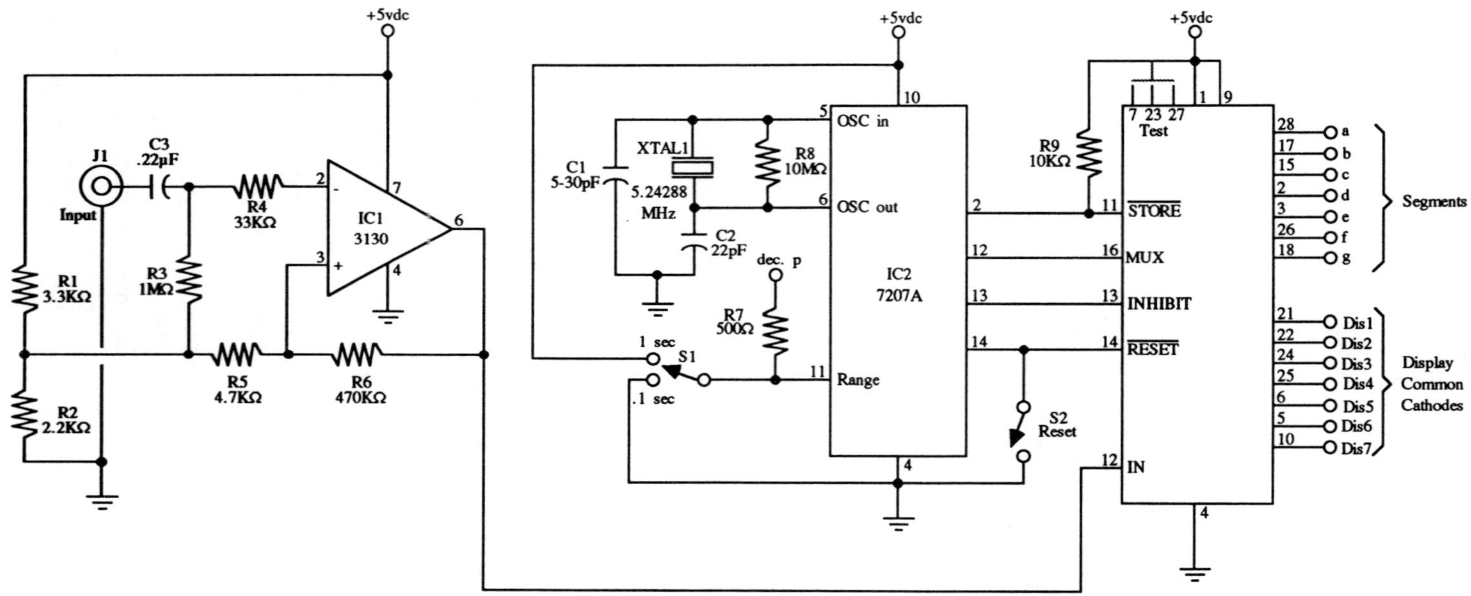

Notes:

Take pin 7, 23, OR 27 HIGH for rapid test of IC3.
Set S1 to 0.1 sec to measure 1Hz to 1MHz.
Set S1 to 1 sec to measure 10 Hz to 10MHz.
Connect R7 to decimal point pin on seven-segmen display.
Depress S2 to reset counter.
Refer to Intersil application notes for additional info on 7207A and 7208 ICs.

Fig. 20-20. Schematic for 10 MHz frequency counter.

Table 20-15. Parts List for 10 MHz Frequency Counter

R1	3.3 kΩ
R2	2.2 kΩ
R3	1 MΩ
R4	33 kΩ
R5	4.7 kΩ
R6	470 kΩ
R7	500 Ω
R8	10 MΩ
R9	10 kΩ potentiometer
C1	5 to 30 pF trimmer cap
C2	22 pF disc capacitor
C3	0.22 μF disc capacitor
IC1	RCA 3130 op amp
IC2	Intersil 7207A time base
IC3	Intersil 7208 frequency counter
XTAL1	5.24288 MHz crystal
S1	SPDT switch
S2	SPST switch
J1	RCA phono jack
DSP1-7	Common-cathode seven-segment display
Misc.	Project box, 5 Vdc power supply

All resistors are 5 to 10 percent tolerance. All capacitors are 10 to 20 percent tolerance, rated at 35 volts or more.

After building the project, test it using the slow-rate variable pulser detailed earlier in this chapter or some other instrument that provides a steady stream of pulses. Connect the pulser to the input of frequency counter. Dial the time base to 1 second, and push the RESET button on the frequency counter. After a moment, the frequency should appear on the display. If nothing happens, or the reading is wildly off, try varying the time-base crystal trimmer capacitor, C1.

When the time base switch is set to 1 second, the frequency range is 1 Hz to 1 MHz. To read higher frequencies, up to 1 MHz, flip the time-base switch to 0.1 second. Remember to multiply the reading by 10 to obtain the actual frequency; for example, if the reading is 100, the actual frequency is 1,000, or 1 kHz.

SPECIAL NOTE: Though this frequency counter is easy to build and use, it does have its limitations. The most serious is that you *cannot* test any signal that applies more than +5 volts to the input because otherwise, the 3130 op amp could be damaged. Also, the companion ICM7207A chip should not be substituted for the older ICM7207. Circuit Specialists (see appendix A) and most other sources that sell the Intersil ICM7208 chip also offer the proper ICM7207A IC as well as the 5.24288 MHz crystal.

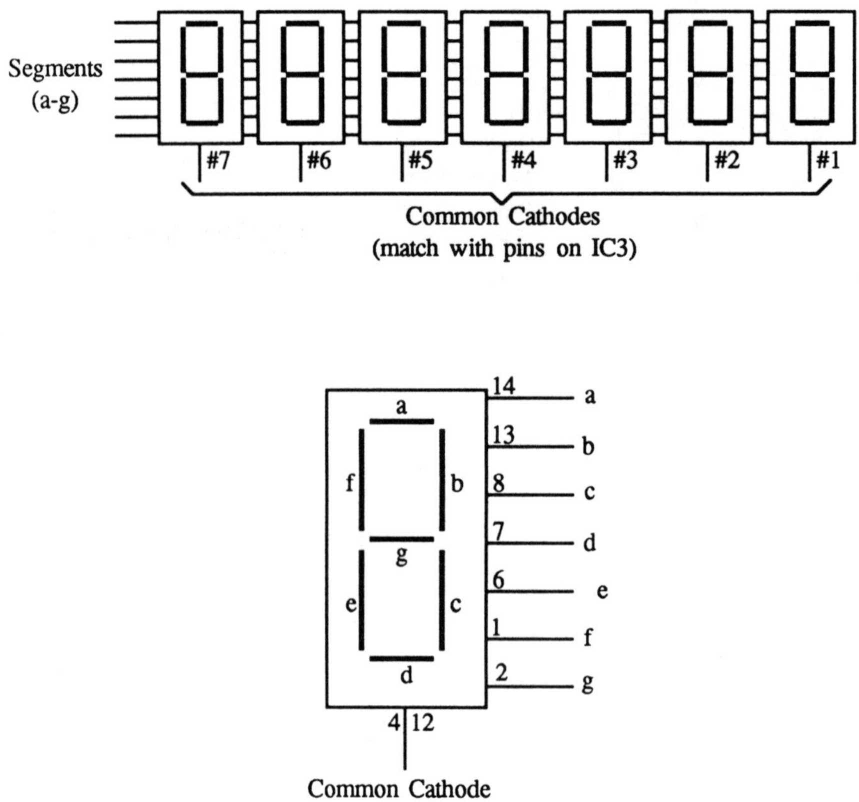

Detail of common cathode
display (pinouts may vary)

Fig. 20-21. How to wire seven-segment LED displays to the 10 MHz frequency counter circuit.

Appendix A
Sources

A.C. INTERFACE, INC.
17911 Sampson Lane
Huntington Beach, CA 92647
High-output LEDs.

ACTIVE ELECTRONICS
P.O. Box 9100
Westborough, MA 01581
(800) ACTIVE-6 (in MA)
(800) ACTIVE-5
Electronic components, test gear, and tools.

ADVANCED FIBEROPTICS CORP.
7650 East Evans Rd, Ste. B
Scottsdale, AZ 85260
(602) 483-7576
Edu-Kit fiberoptic kit, fiberoptic components, and systems.

ALL ELECTRONICS CORP.
P.O. Box 567
Van Nuys, CA 91408
(800) 826-5432
(818) 904-0524 (in CA)

Retail and surplus components, switches, relays, keyboards, transformers, computer-grade capacitors, cassette player/recorder mechanisms, etc. Mail-order and retail stores in L.A. area. Regular catalog.

ALLEGRO ELECTRONICS SYSTEMS
3E Mine Mountain
Cornwall Bridge, CT 06754
(203) 672-0123
Mail order. Laser and high-voltage equipment. Catalog available.

ALLIED ELECTRONICS
401 E. 8th St.
Ft. Worth, TX 76102
(800) 433-5700
Electronic parts outlet; catalog and regional sales offices. Good source for hard-to-find items.

ALLKIT ELECTRONICS
434 W. 4th St.
West Islip, NY 11795
Electronic components, grab bags, and switches—all at attractive prices.

ALLTRONICS
15460 Union Ave.
San Jose, CA 95124
(408) 371-3053
Mail order and store in San Jose, CA. New and surplus electronics.

ALPHA PRODUCTS
7904-N Jamaica Ave.
Woodhaven, NY 11421
(800) 221-0916 (orders)
(203) 656-1806 (info)
(718) 296-5916 (NY orders)
Mail order. Stepper motors and stepper motor controllers (ICs and complete boards), process control devices (relay cards, touch-tone decoders, etc.).

AMERICAN SCIENCE CENTER, INC.
601 Linden Place
Evanston, IL 60602
(312) 475-8440
Parent company to surplus outfit Jerryco, the American Science Center sells new

and surplus lab and optic components. Most of the line is drawn from Edmund Scientific stock at about the same prices.

AMERICAN DESIGN COMPONENTS
815 Fairview Ave.
P.O. Box 220
Fairview, NJ 07022
(201) 941-5000
New and surplus electronics, motors, and computer gear.

AMICOM
25 Hartwell Ave.
Lexington, MA 02173
Conductive ink pens.

ANALYTIC METHODS
1800 Bloomsbury Ave.
Ocean, NJ 07712
(201) 922-6663
Mail order. Integrated circuits, LEDs, surplus parts, computer cables.

ANCHOR ELECTRONICS
2040 Walsh Ave.
Santa Clara, CA 95050
(408) 727-3693
Electronic components. Catalog available.

ANTIQUE ELECTRONIC SUPPLY
688 W. First St.
Tempe, AZ 85281
Parts and supplies for antique radios; good source for hard-to-find items like high-voltage capacitors and large-wattage resistors.

BARRETT ELECTRONICS
5312 Buckner Dr.
Lewisville, TX 75028
Mail-order electronic surplus: components, power supplies, computer-grade capacitors.

BCD ELECTRO
P.O. Box 830119
Richardson, TX 75083-0119

Electronic components of all types. Some surplus, including laser systems. Very good prices. Catalog available.

BIGELOW ELECTRONICS
P.O. Box 125
Bluffton, OH 45817-0125
New and surplus electronics, hardware, and ham gear. Catalog available.

BISHOP GRAPHICS
5388 Sterling Center Dr.
P.O. Box 5007EZ
Westlake Village, CA 91359
(818) 991-2600
EZ Circuit PCB-making supplies.

C & H SALES
2176 E. Colorado Blvd.
Pasadena, CA 91107
(800) 325-9465
Electronic and mechanical surplus including some optics like prisms, mirrors, and lenses. Reasonable prices for most items; regular catalog. Their walk-in store in Pasadena has more items than those listed in the catalog, so if you're in the area, be sure to drop in.

CIRCUIT SPECIALISTS
Box 3047
Scottsdale, AZ 85257
(602) 966-0764
New electronic goodies, including Intersil frequency counter chips. Full complement of standard components at good prices.

COMPUTER SURPLUS STORE
715 Sycamore Dr.
Milpitas, CA 95035
(408) 434-0168
Surplus electronics. Mostly computers, printers, and power supplies, but also carries components. Walk-in store in Northern California and limited mail order.

COMPUTER PARTS MART
3200 Park Blvd.
Palo Alto, CA 94306
(415) 493-5930
Mail-order surplus. Regular catalog.

CONSOLIDATED ELECTRONICS INC.
705 Watervliet Ave.
Dayton, OH 45420-2599
General components.

DATAK CORP.
3117 Paterson Plank Rd.
North Bergen, NJ 07047
(201) 863-7667
Supplies for making your own printed circuit boards including the popular direct-etch dry-transfer method.

DEXTER CORP
Hyson Division
15051 E. Don Julian Rd.
Industry City, CA 91749
Conductive ink pens.

DICK SMITH ELECTRONICS, USA
P.O. Box 468
Greenwood, IN 46142
General electronic components.

DIGI-KEY CORP.
701 Brooks Ave. South
P.O. Box 677
Thief River Falls, MN 56701-0677
(800) 344-4539
Mail-order. Discount components—everything from crystals to integrated circuits to resistors and capacitors in bulk. Catalog available.

DOKAY COMPUTER PRODUCTS
2100 De la Cruz Blvd.
Santa Clara, CA 95050
(408) 988-0697
(800) 538-8800; (800) 848-8008 (in CA)
Electronic parts (most popular linear, TTL, and CMOS devices) and computer gear (including computer chips). Catalog available.

DOLAN-JENNER INDUSTRIES, INC.
P.O. Box 1020
Woburn, MA 01801
Fiberoptic components.

DYNALOW INC.
7 Great Meadow Lane
Hanover, NJ 07936
Conductive ink pens.

EDLIE ELECTRONICS
2800 Hempstead Turnpike
Levittown, NY 11756
(800) 645-4722
(516) 735-3330
Electronic components, test gear, surplus components.

EDMUND SCIENTIFIC CO.
101 E. Gloucester Pike
Barrington, NJ 08007-1380
(609) 573-6250
Mail order. New and surplus motors, gadgets, and other goodies. Regular catalog (ask for both the Hobby and Industrial versions).

ELECTRONIC LIQUIDATORS
P.O. Box 27656
Lansing, MI 48901
Electronic surplus.

ELECTROVALUE INTERNATIONAL
Box 376
Morris Plains, NJ 07950
(201) 267-1117
Electronic components.

ERAC CO.
8280 Clairemont Mesa Blvd., Suite 117
San Diego, CA 92111
(619) 569-1864
Mail order and retail store. PC-compatible boards and subsystems, surplus computer boards, computer power supplies and components.

FAIR RADIO SALES
1016 E. Eureka St.
P.O. Box 1105
Lima, OH 45802
(419) 223-9176; (419) 227-6573
An old and established retailier of surplus goods of all types, particularly military

communications. A boon to the amateur radio enthusiast and moderately useful to the general electronics experimenter. Catalog available.

FOBTRON COMPONENTS
21738 S. Avalon Blvd.
Suite #155
Carson, CA 90745
Electronic and laser components. Flyer available.

FORDHAM RADIO
260 Motor Parkway
Hauppauge, NY 11788
(800) 645-9518
(516) 435-8080 (in NY)
Electronic test equipment, tools, power supplies. Catalog available.

G.B. MICRO
P.O. Box 280298
Dallas, TX 75228
(214) 271-5546
Mail order. Components including 3.12 MHz crystals, 300 kHz crystals, UARTs, the most popular TTL, CMOS, and linear ICs, and construction parts.

GENERAL SCIENCE & ENGINEERING
P.O. Box 447
Rochester, NY 14603
(716) 338-7001
Surplus electronics; some kits (laser power supply, variable-rate electronic strobe, light-beam communicator, and more).

GIANT ELECTRONICS, INC.
19 Freeman St.
Newark, NJ 07105
(800) 645-9060
(201) 344-5700
New and surplus electronics, motors, computer peripherals, and more.

GIL ELECTRONICS
P.O. Box 1628
911 Hidden Valley
Soquel, CA 95073
Electronic parts, books, new and surplus computer components.

H&R CORP.
401 E. Erie Ave.
Philadelphia, PA 19134
(215) 426-1708
Surplus mechanical and electronic components. Excellent source for heavy-duty DC gear motors. Regular catalog.

HAL-TRONIX INC.
12671 Dix-Toledo Highway
P.O. Box 1101
Southgate, MI 48195
(313) 281-7773
Mail order. Surplus computers, computer components, PC-compatible boards and subsystems.

HALTED SPECIALITIES CO.
3500 Ryder St.
Santa Clara, CA 95051
(408) 732-1573
Mail order and retail stores. New and surplus components, PC-compatible boards and subsystems, power supplies, most popular ICs, transistors, etc.

HEATHKIT
P.O. Box 1288
Benton Harbor, MI 49022
(800) 253-0570
Perhaps the best source of professionally produced electronic kits. Also available are testing gear (assembled and kit form) and computers.

HOSFELT ELECTRONICS, INC.
2700 Sunset Blvd.
Stubenville, OH 43952
(800) 524-6464
New and surplus electronics at attractive prices. Catalog available.

INFORMATION UNLIMITED
Box 716
Amherst, NH 03031
(603) 673-4730
Plans and kits for lasers and other high-tech gadgets. Good source for reasonably priced new laser tubes.

J.I. MORRIS CO.
394 Elm St.
Southbridge, MA 01550
(617) 764-4394
Small components and hardware; miniature screws, taps, and nuts/bolts.

JAMECO ELECTRONICS
1355 Shoreway Rd.
Belmont, CA 94002
(415) 592-8097
Mail order. Components, PC-compatible boards and subsystems. Regular catalog.

JDR MICRODEVICES
1224 S. Bascom Ave.
San Jose, CA 95128
(408) 995-5430
Large selection of new components, wire-wrap supplies, PC-compatible boards and subsystems. Mail order and retail stores in San Jose area.

JERRYCO, INC.
601 Linden Place
Evanson, IL 60202
(312) 475-8440
Regular catalog lists hundreds of surplus mechanical and electronic gadgets for robots. Good source for motors, rechargeable batteries, switches, solenoids, and lots more.

JOHN J. MESHNA, JR. CO.
P.O. Box 62
E. Lynn, MA 01904
(617) 595-2272
New and surplus merchandise. Catalog available.

LINDSAY PUBLICATIONS
P.O. Box 12
Bradley, IL 60915
Books on electrical and mechanical design, much of it dating back to the early part of the 20th century. Good for history.

MARK V ELECTRONICS
8919 E. Slauson Ave.
Montebello, CA 90640
Electronic kits.

MARTIN P. JONES & ASSOC.
P.O. Box 12685
Lake Park, FL 33403-0685
(407) 848-8236
All sorts of goodies including electronic components, (transistors, capacitors, resistors, ICs, etc.), optical stuff, and more. Catalog available.

MCM ELECTRONICS
858 E. Congress Park Dr.
Centerville, OH 45459-4072
(513) 434-0031
Test equipment, tools, and supplies. Catalog available.

MEADOWLAKE
25 Blanchard Dr.
Northport, NY 11768
TEC-200 direct-transfer film for making quick integrated circuits from artwork in magazines and this book. All you need is a plain-paper copier (or access to one).

MEREDITH INSTRUMENTS
P.O. Box 1724
Glendale, AZ 85311
(602) 934-9387
Laser surplus including tubes, power supply, optics, and components. Stock comes and goes, so call first to make sure they have what you want. Flyer available; very good prices.

MICRO MART, INC.
508 Central Ave.
Westfield, NJ 07090
(201) 654-6008
Electronic components (ICs, transistors, etc.). Their grab bags are better than average, and prices are low.

MOUSER ELECTRONICS/TEXAS DISTRIBUTION CENTER
2401 Hwy. 287 North
North Mansfield, TX 76063
(817) 483-4422
Mail order. Discount electronic components. Catalog available.

NEWARK ELECTRONICS
4801 North Ravenswood Ave.
Chicago, IL 60640
(312) 784-5100
Authorized dealer for many lines of electronic component manufacturers. Catalog available; regional offices.

OCTE ELECTRONICS
Box 276
Alburg, VT 05440
(514) 739-9328
New and surplus electronic stuff at reasonable prices. OCTE carries a lot of complete components, such as cable converters and power supplies, but individual parts, like batteries and connectors, are also usually available.

PARTS EXPRESS INTERNATIONAL, INC.
340 East First St.
Dayton, OH 45402
(800) 338-0531
Repair parts (TVs, VCRs, etc.). Catalog available.

PLANNED PRODUCTS
303 Potrero St.
Suite 53
Santa Cruz, CA 95060
Conductive ink pens.

PRECISION ELECTRONICS CORP.
605 Chestnut St.
Union, NJ 07083
(800) 255-8868
(201) 686-4646 (in NJ)
Electronic components—linear ICs, transistors, hard-to-find Sanyo and Zenith ICs (mostly for TV), and more.

R&D ELECTRONIC SUPPLY
100 E. Orangethorpe Ave.
Anaheim, CA 92801
(714) 773-0240
Mail order. Power supplies, computer equipment, test equipment.

R&D ELECTRONICS
1202H Pine Island Rd.
Cape Coral, FL 33909
(813) 772-1441
Mail order. New and surplus electronics; components, switches, ICs.

RADIO SHACK
One Tandy Center
Fort Worth, TX 76102
Nation's largest electronics retailer. Many popular components, though short on ICs. Good source for general electronic needs, including some fiberoptic components. Catalog available through store.

SHARON INDUSTRIES
1919 Hartog Road
San Jose, CA 95131
(408) 436-0455
Mail order and retail store. New and surplus electronic components, ICs, computers, PC-compatible boards and subsystems.

SILICON VALLEY SURPLUS
4401 Oakport
Oakland, CA 94601
(415) 261-4506
Surplus electronic, computer, and mechanical goodies. Mail order, and retail storing Oakland, CA.

SINTEC COMPANY
28 8th St.
P.O. Box 410
Frenchtown, NJ 08825
(800) 526-5690
(201) 996-4093 (NJ)
New electronic components. Catalog available.

SMALL PARTS
6901 NE Third Ave.
Miami, FL 33238
(305) 751-0856
A potpourri of small parts, ideally suited for miniature mechanisms. Not cheap, but good quality.

Solid State Sales
P.O. Box 74D
Somerville, MA 02143
(800) 343-5230
(617) 547-7053
New electronic components. Catalog available.

Stock Drive Products
55 S. Denton Ave.
New Hyde Park
New York, NY 11040
Assortment of mechanical components, including gears, shafts, and motors.

Synergetics
Box 809
Thatcher, AZ 85552
(602) 428-4073
Columnist's Don Lancaster's company, with "hacker's help line." Lancaster offers free technical advice (within limits), a collection of his old and new books (his TTL and CMOS cookbooks are classics), and several "info packs" of useful programming, graphics, and technical stuff.

Unicorn Electronics
10010 Canoga Ave., Unit B-8
Chatsworth, CA 91311
(800) 824-3432
(818) 341-8833
All popular electronic parts including resistors, capacitors, ICs, and components. Some surplus. All at good prices.

United Products, Inc.
1123 Valley Rd.
Seattle, WA 98109
(206) 682-5025
Mail order. Stepper motors, computer components, test equipment.

Windsor Distributors
19 Freeman St.
Newark, NJ 07105
(800) 645-9060
Mail order. Surplus electronics.

YEARY COMMUNICATIONS
12922 Harbor Blvd.
Suite 800
Garden Grove, CA 92640
General electronic components, including radio.

Appendix B
Further reading

Here is a selected list of magazines and books that can enrich your understanding and enjoyment of all facets of electronics.

Magazines

Computer Shopper
5211 S. Washington Ave.
P.O. Box F
Titusville, FL 32781
Monthly magazine (some call it a bible) for computer enthusiasts. Probably more ads than articles, Computer Shopper *carries updated listings of swap meets and bulletin boards as well as classified advertising from small surplus dealers.*

Electronics Today
1300 Don Mills Rd.
Toronto, Ontario M3B 3M8
Canada
General-interest electronics magazine with emphasis on hobby how-to articles.

Hands-On Electronics
500 Bi-County Blvd.
Farmingdale, NY 11735

Monthly magazine put out by the editors of Radio-Electronics. *The articles and construction projects are aimed at beginning electronics enthusiasts.*

Modern Electronics
76 North Broadway
Hicksville, NY 11801
Monthly magazine for electronics hobbyists. Don't miss the regular columns by hobby electronic guru Forest Mims III. Many of the editors used to be involved with Popular Electronics *before that magazine changed over to computer-only coverage (and then ceased publication). Check back issues of* Modern Electronics *for Don Lancaster's fascinating columns (he moved his "Hardware Hacker" column to* Radio-Electronics *in late 1987).*

Nuts & Volts
P.O. Box 1111
Placentia, CA 92670
Monthly "magazine" that contains only advertising—both display ads and classifieds. Often carries ads for new and used electronic equipment and lists upcoming swap meets (both computer and electronic). Subscription price is fairly low.

QST
225 Main St.
Newington, CT 06111
Monthly magazine aimed at the amateur radio enthusiast, but it often carries articles of interest to the general electronics experimenter.

Radio-Electronics
500 Bi-County Blvd.
Farmingdale, NY 11735
Monthly magazine for electronics hobbyists. Be sure to read the column by Don Lancaster of Synergetics. His tips and hands-on help are invaluable for general electronics experimentation. He also provides sources for hard-to-find parts and information.

Books

Basic Digital Electronics—2nd Edition
Ray Ryan and Lisa A. Doyle; TAB Books, catalog #3370.
Introduction to the principles of digital theory.

Beginner's Guide to Reading Schematics—2nd Edition
Robert J. Traister and Anna L. Lisk; TAB Books, catalog #3632
How to read and interpret schematic diagrams.

Build Your Own Laser, Phaser, Ion Ray Gun & Other Working Space-Age Projects
Robert E. Iannini; TAB Books, catalog #1604
A guide to making high-tech gadgets, with some insight and designs for laser-based projects. Iannini, who runs a mail-order firm, presents six laser projects plus a few others (such as infrared light detection) that can be used with laser systems. A companion book, Build Your Own Working Fiberoptic, Infrared and Laser Space-Age Projects *(catalog #2724) offers more laser-based designs.*

Circuit Scrapbook
Forrest Mims III; McGraw-Hill, Inc.
More of Mims' Popular Electronics *columns—these from 1979 to 1981. Several good circuits and designs that can be adapted for laser work. Also check out the sequel* Circuit Scrapbook II *(Howard W. Sams Co.).*

CMOS Cookbook
Don Lancaster; Howard W. Sams Co.
A classic in its own time, the CMOS Cookbook *presents useful design theory and practical circuits for many popular CMOS chips. The companion book,* TTL Cookbook, *is equally as helpful.*

Engineer's Mini-Notebook
Forrest Mims III; Radio Shack book series
The Engineer's Mini-Notebooks *is a series of small books written by Forrest Mims that cover a wide variety of hobby electronics: from using the NE555 timer to optoelectronics circuits to op-amp circuits, and more. The entire set is a "must have," and besides, they're cheap.*

44 Power Supplies for Your Electronic Projects
Traister and Mayo; TAB Books, catalog #2922
Forty-four complete power supplies designed for general electronics projects.

Handbook of Microcomputer Interfacing
Steve Leibson; TAB Books, catalog #3101
How to connect outside devices and circuits to a computer or microprocessor.

Troubleshooting and Repairing Electronic Circuits
Robert L. Goodman; TAB Books, catalog #3258
General troubleshooters guide for both analog and digital.

Laser Cookbook
Gordon McComb; TAB Books, catalog #3090
An inside look at helium-neon and solid-state lasers, plus numerous projects you

can build including light show systems, a holographic sandbox, and a handheld laser pistol.

Programmer's Problem Solver, for the IBM PC, XT, & AT
Robert Jourdain; Brady
A technical book on the inner workings of the IBM PC with special emphasis on programming in BASIC, assembly, and machine code. Extensive section on parallel ports.

Robot Builder's Bonanza
Gordon McComb; TAB Books, catalog #2800
Book on a compendium of projects useful to the robot experimenter. Some of the information, such as dc motor operation and remote control, can be suitably adapted to many kinds of hobby electronics projects.

Silliconnections: Coming of Age in the Electronic Era
Forrest M. Mims III; McGraw-Hill, Inc.
A history and personal autobiography of the electronics revolution. Mims, a writer and inventor, contributed to the development of the first personal computer and was employed at the Air Force Weapons Laboratory in New Mexico where he worked with early solid-state diodes.

Index

D

E